# Social and Ecological Interactions in the Galapagos Islands

T0093422

For further volumes:
http://www.springer.com/series/10427

Stephen J. Walsh · Carlos F. Mena
Editors

# Science and Conservation in the Galapagos Islands

## Frameworks & Perspectives

*Editors*
Stephen J. Walsh
Department of Geography
University of North Carolina
  at Chapel Hill
Chapel Hill, NC, USA

Carlos F. Mena
College of Biological and Environmental
  Sciences
Universidad San Francisco de Quito
Quito, Ecuador

ISSN 2195-1055            ISSN 2195-1063 (electronic)
ISBN 978-1-4614-5793-0    ISBN 978-1-4614-5794-7 (eBook)
DOI 10.1007/978-1-4614-5794-7
Springer New York Heidelberg Dordrecht London

Library of Congress Control Number: 2012953606

# Foreword

## Introduction to the Galapagos Initiative

On May 16, 2011, the Galapagos Science Center (GSC), on San Cristobal Island in the Galapagos Archipelago of Ecuador, was dedicated by delegations led by Chancellor Santiago Gangotena, Universidad San Francisco de Quito (USFQ), Ecuador, and Chancellor Holden Thorp, University of North Carolina at Chapel Hill (UNC), USA. The dedication marked a significant and conspicuous point in the USFQ–UNC relationship, initially developed around the Galapagos Initiative—an interdisciplinary program of research, education, and community outreach and engagement that examines the social, terrestrial, and marine subsystems of the Galapagos Islands. Scientists from the social, natural, spatial, and computational sciences have come together from both campuses, as well as from international institutions, to participate in the Galapagos Initiative, developed and led by Carlos F. Mena at USFQ and Stephen J. Walsh at UNC, codirectors of the Galapagos Science Center. With the full support of their respective campuses, USFQ and UNC have combined talents and expertise in a unique partnership to create the Galapagos Science Center and to address the social and ecological sustainability of the Galapagos Islands.

The Galapagos Science Center is equipped with four research and education laboratories: Microbiology, Marine Ecology, Terrestrial Ecology, and Spatial Analysis and Modeling. In addition, the facility includes faculty, student, and staff offices, a conference room, a community classroom, and associated space for equipment, experiments, and education and community outreach programs. The Galapagos Science Center is also equipped with a sophisticated cyber infrastructure, staffed with full-time professionals, and linked to the local community and the archipelago more generally.

Through Study Abroad programs, UNC also is linking undergraduate students to the Galapagos Science Center, working through the USFQ Galapagos Academic Institute for Arts and Sciences (GAIAS), directed by Diego Quiroga and Carlos Valle. In additional several master's theses and doctoral dissertations have been completed through the Galapagos Initiative on an array of topics including local fisheries, institutions and policies, mangroves, and land use and tourism. Several others are underway. The Galapagos Science Center has also attracted a diverse and talented group of research faculty and educators from the respective campuses who are addressing collaborative topics that extend across the sciences and resonate throughout the Galapagos Islands, and well beyond. In addition, projects are being conducted in close association with the Galapagos National Park, institutes and agencies of the Ecuadorian government, and several NGOs in the Galapagos Islands.

## Introduction to the Series and the Book to Launch the Series

Motivated to understand the social, terrestrial, and marine subsystems of the Galapagos Island and their social–ecological interactions, we have created a book series titled *Social and Ecological Interactions in the Galapagos Islands* that will be published by Springer Science and Business Media, beginning with this initial book. We anticipate the publication of at least one volume each year on an array of topics that collectively examine science, conservation, population, health, and environment in the Galapagos Islands. This initial book to launch the series is titled *Science and Conservation in the Galapagos Islands: Frameworks and Perspectives*. It is designed to provide a broad description of the many challenges and opportunities for examining the Galapagos Islands through multiple lenses that offer a freshness of view and perspective, where the human dimension is clearly linked to the environment through the perspective that we call *"Island Biocomplexity."* *Island Biocomplexity* combines social–ecological coevolution and adaptive resilience with a new island ecology that incorporates human impacts in coupled natural-human systems. *Island Biocomplexity* also encompasses the interactions within and among ecological systems, the physical systems on which they depend, and the human systems with which they interact (Michener et al. 2003; Walsh et al. 2011).

This theoretical approach accommodates an enhanced understanding of the linked and integrative effects of people and environment and the nature of feedbacks that create nonlinear system dynamics, improved pattern–process understanding, and explicit linkages between the social, terrestrial, and marine subsystems of the Galapagos Islands, achieved through an integrated systems perspective (Gonzalez et al. 2008; Miller et al. 2010; Mena et al. 2011). This initial book creates a broad foundation for the more focused volumes to follow that will be developed by guest editors, responding to questions that resonate in science, conservation, and society.

In this book, we draw together scholars, predominately from the University of North Carolina at Chapel Hill and the Universidad San Francisco de Quito, as well as scientists from collaborating institutions. Collectively, these authors describe conditions, challenges, and changes in the Galapagos Islands as viewed through multiple frameworks and perspectives that include Evolutionary Biology, Complexity Theory and Agent-Based Models, Political Ecology, Landscape Ecology, Sustainability Science, Marine Conservation, Invasive Species, Water Quality, Household Livelihoods, Land Use Dynamics, Tourism, Political Institutions, Nutrition and Health, and History as context for the Galapagos

Islands as a special and contested place. Future book topics will be shaped by the interests and ideas developed by the coeditors and by the series editorial board, listed below:

## Series Editorial Board

Stephen J. Walsh and Carlos F. Mena, Series Editors
Margaret Bentley, University of North Carolina at Chapel Hill, USA
Eliecer Cruz, World Wildlife Fund, Galapagos Islands, Ecuador
Joel Fodrie, University of North Carolina at Chapel Hill, USA
Judith Denkinger, Universidad San Francisco de Quito, Ecuador
Jonathan Lees, University of North Carolina at Chapel Hill, USA
Kenneth Lohmann, University of North Carolina at Chapel Hill, USA
Aaron Moody, University of North Carolina at Chapel Hill, USA
Diego Quiroga, Universidad San Francisco de Quito, Ecuador
Gunter Reck, Universidad San Francisco de Quito, Ecuador
Ronald Rindfuss, University of North Carolina at Chapel Hill, USA
Conghe Song, University of North Carolina at Chapel Hill, USA
Stella de la Torre, Universidad San Francisco de Quito, Ecuador
Maria de Lourdes Torres, Universidad San Francisco de Quito, Ecuador
Gabriel Trueba, Universidad San Francisco de Quito, Ecuador
Gabriela Valdivia, University of North Carolina at Chapel Hill, USA
Carlos Valle, Universidad San Francisco de Quito, Ecuador
Matthias Wolff, University of Bremen, Germany

Chapel Hill, NC, USA                                                                    Stephen J. Walsh
Quito, Ecuador                                                                             Carlos F. Mena

# References

Gonzalez JA., Montes C, Rodríguez J, Tapia W (2008). Rethinking the Galapagos Islands as a complex social–ecological system: implications for conservation and management. Ecol Soc 13(2): 13 (online)

Mena CF, Walsh SJ, Frizzelle BG, Malanson GP (2011). Land use change of household farms in the Ecuadorian Amazon: design and implementation of an agent based model. Appl Geogr 31(1):210–222

Michener WK, Baerwald TJ, Firth P, Palmer MA, Rosenberger J, Sandlin EA, Zimmerman H (2003) Defining and unraveling biocomplexity, Bioscience. 51(12), 1018–1023

Miller BW, Breckheimer I, McCleary AL, Guzman-Ramirez L, Caplow SC, Walsh SJ (2010) Using stylized agent-based models for population-environment research: a case from the Galapagos Islands. Popul Environ 31(6):401–426

Walsh SJ, Malanson GP, Messina JP, Brown DG, Mena CF (2011) Biocomplexity. In: Millington A, Blumler M, Schickhoff U(Eds.) The SAGE handbook of biogeography. SAGE Publications, London, p 469–487

# Preface

The Galapagos Archipelago has been the source of inspiration to many individuals, and over the human history of the archipelago, many different views and opinions have been the source of debate around the world. I will not try to match those brilliant minds who have done great jobs in portraying Galapagos for the benefit of mankind, but I will try to tell my story and my history, and perhaps through my eyes the reader will feel the same passion that has driven me and forged my efforts to contribute to the maintenance and improvement of the conservation of this unique archipelago, a place that I have the honor to call home.

The authors of the chapters in this book are a perfect example of that passion that Galapagos inspires in all of us. In each chapter there is a tremendous amount of research, which is the result of dedicated passion for Galapagos, and perhaps together we might find a way to better understand the concept of sustainability that is so badly needed worldwide.

I was born on Floreana Island, of the four inhabited ones, the smaller island, both geographically and in the number of people living there. I believe I was born with an interest in the natural history of the islands, not in a conscious way, but very interested in all terrestrial life forms that occur on the island. Since my childhood, I have witnessed great changes in the islands; positive ones and some others not so positive. But one constant element over the years has been the tremendous resilience of the native and endemic species to cope with those changes—from feeding habits, behavioral choices to settle on new habitats, ability to cope with the arrival of new species, the street lights, traffic noise, speed boats, oil spills, and to the many visitors who constantly arrive to see the natural wonders that apparently inspired Charles Darwin in the thinking of the theory of evolution.

Also, another constant element that I have witnessed is the status of conservation of the Galapagos Islands that gets better each year. The Galapagos are one of the last few well-conserved tropical archipelagos in the world, which is home to numerous endemic species, and since Darwin's 1835 visit, the archipelago has been the focus of critical research on natural selection and evolution. This extraordinary archipelago has inspired countless visitors, and I'm not going to try to rewrite any of those descriptions, but instead narrate the changes that I have witnessed during my life, which in turn relates to the well-researched chapters that the different authors have put together in this book. Some of those chapters are the results of research that has been overlooked in the past; needed research, indeed, if we are to manage Galapagos as a socio-ecosystem, which has been the case since 1832 when Ecuador took possession of the islands. Since then, the islands have gone through a series of changes that have shaped the present, and I'm sure, the future of the archipelago.

Galapagos was one of the first World Heritage Sites inscribed by UNESCO in 1978, because of the archipelago's outstanding universal value and its integrity as a complete ecosystem that can ensure ecological and evolutionary processes. World Heritage recognition was extended to the Galapagos Marine Reserve in 2001, because of its endemic diversity, multifaceted oceanic currents, unusual biogeography, and complex marine communities that are linked to the terrestrial ecosystems.

The Galapagos Archipelago has been in the eye of the world due to its unique endemic species and, in the last century, the world has come to believe that Galapagos will not survive, thus sending expeditions to collect as many samples as possible of the endemic species to save them in collections cabinets for future generations. So far, the status of conservation of the archipelago is better than a century ago. But it is also true that the threats that the Galapagos faces are greater than before.

Islands are ecologically, economically, politically, and socially different from the rest of the world. In reality, islands should be kept "at risk" forever due to the loss of the original isolation that is the basis for the existence of the endemic species as well as those evolutionary processes that are so strong in isolated islands. We (*homo sapiens*) have been able to break those natural barriers. Today, the Galapagos Islands are visited by several planes each day, and also by several cargo boats each month. In each of those events, as it has been in the past, new species arrive that are aided by the easy transportation to the islands. And, there is an ever-increasing risk that with each plane or each boat, new invasive species could arrive and settle in the archipelago. The effects of globalization are not always good and, in this case, it could be the tipping point for the future of Galapagos.

I have been able to witness changes in Galapagos during my life. Since I'm one of the fortunate ones to have the honor to be a "native" growing up and working in what I call home, it has allowed me to observe and be part of a series of positive and negative changes. I'll describe some of them, but in reality each chapter in this book is a sample of how changes are occurring in the Galapagos and their possible meaning to the Galapagos.

I have experienced a life of change in the Galapagos: from no electricity to cable TV, through eventual contacts with the outside world, to now over 150,000 visitors each year; from a mail system that took several months if not years for letters and packages to arrive, to Internet and e-mail systems; from the happiness to know that Lonesome George was found, to his death as I write these words; and from the discovery of the pink iguana, a large vertebrate that is totally new to science. The islands have gone from introducing herbivores to eradicating them, as one of the most successful projects worldwide; from elementary schools to universities; and from importing skills to exporting advice in many areas and to many parts of the world. These changes show that Galapagos is a growing community aware of its unique environment and that Ecuador and the world are still keeping an interest in and care for this unique archipelago.

I truly hope that this series of publications will help to disseminate the challenges that Galapagos faces and will be a great tool to the decision-making process to have a better Galapagos for future generations. The death of Lonesome George should remind us that extinctions are forever, and, therefore, we should be inspired to do all that we can and more to prevent future losses of unique species.

Galapagos, Ecuador                                                                                    Felipe Cruz

# Contents

1  Science and Conservation in the Galapagos Islands............................  1
   Carlos A. Valle

2  Changing Views of the Galapagos ...........................................  23
   Diego Quiroga

3  Perspectives for the Study of the Galapagos Islands:
   Complex Systems and Human–Environment Interactions...............  49
   Stephen J. Walsh and Carlos F. Mena

4  The Socioeconomic Paradox of Galapagos.........................................  69
   Byron Villacis and Daniela Carrillo

5  Environmental Crisis and the Production of Alternatives:
   Conservation Practice(s) in the Galapagos Islands............................  87
   Wendy Wolford, Flora Lu, and Gabriela Valdivia

6  The Double Bind of Tourism in Galapagos Society ...........................  105
   Laura Brewington

7  The Evolution of Ecotourism: The Story of the
   Galapagos Islands and the Special Law of 1998 ...............................  127
   Michele M. Hoyman and Jamie R. McCall

8  People Live Here: Maternal and Child Health
   on Isla Isabela, Galapagos......................................................  141
   Rachel Page, Margaret Bentley, and Julee Waldrop

9  Characterizing Contemporary Land Use/Cover
   Change on Isabela Island, Galápagos .............................................  155
   Amy L. McCleary

10 Investigating the Coastal Water Quality
   of the Galapagos Islands, Ecuador......................................................  173
   Curtis H. Stumpf, Raul A. Gonzalez, and Rachel T. Noble

11  Research in Agricultural and Urban Areas
    in Galapagos: A Biological Perspective ............................................... 185
    Stella de la Torre

12  A Geographical Approach to Optimization
    of Response to Invasive Species ............................................................ 199
    George P. Malanson and Stephen J. Walsh

13  From Whaling to Whale Watching: Cetacean Presence
    and Species Diversity in the Galapagos Marine Reserve ................... 217
    Judith Denkinger, Javier Oña, Daniela Alarcón,
    Godfrey Merlen, Sandy Salazar, and Daniel M. Palacios

Index ................................................................................................................ 237

# Contributors

**Daniela Alarcón** Universidad San Francisco de Quito, Quito, Ecuador

**Margaret Bentley** Gillings School of Global Public Health, University of North Carolina at Chapel Hill, Chapel Hill, NC, USA

**Laura Brewington** Department of Geography, University of North Carolina, Chapel Hill, NC, USA

**Daniela Carrillo** National Institute of Statistics and Census – INEC, Quito, Ecuador

**Felipe Cruz** Native Naturalist, Puerto Ayora, Santa Cruz Island, Galapagos, Ecuador

**Stella de la Torre** College of Biological and Environmental Sciences, Universidad San Francisco de Quito, Quito, Ecuador

**Judith Denkinger** Galapagos Science Center, College of Biological and Environmental Sciences, Universidad San Francisco de Quito, Quito, Ecuador

**Raul A. Gonzalez** Institute of Marine Sciences, University of North Carolina at Chapel Hill, Morehead City, NC, USA

**Michele M. Hoyman** Department of Political Science, University of North Carolina at Chapel Hill, Chapel Hill, NC, USA

**Flora Lu** Department of Environmental Studies, University of California, Santa Cruz, CA, USA

**George P. Malanson** Department of Geography, University of Iowa, Iowa City, IA, USA

**Jamie R. McCall** Department of Public Administration, North Carolina State University, Raleigh, NC, USA

**Amy L. McCleary** Department of Geography, Center for Galápagos Studies, University of North Carolina at Chapel Hill, Chapel Hill, NC, USA

**Carlos F. Mena** College of Biological and Environmental Sciences, Galapagos Science Center, University of San Francisco, Quito, Ecuador

**Godfrey Merlen** Galapagos National Park, Puerto Ayora, Santa Cruz, Ecuador

**Rachel T. Noble** Institute of Marine Sciences, University of North Carolina at Chapel Hill, Morehead City, NC, USA

**Javier Oña** Universidad San Francisco de Quito, Quito, Ecuador

**Rachel Page** Gillings School of Global Public Health, University of North Carolina at Chapel Hill, Chapel Hill, NC, USA

**Daniel M. Palacios** Joint Institute for Marine and Atmospheric Research, University of Hawaii, Honolulu, HI, USA

NOAA/NMFS/SWFSC/Environment Research Division, Pacific Grove, CA, USA

**Diego Quiroga** College of Biological and Environmental Sciences, Universidad San Francisco de Quito, Quito, Ecuador

**Sandy Salazar** Joint Institute for Marine and Atmospheric Research, University of Hawaii, Honolulu, HI, USA

NOAA/NMFS/SWFSC/Environment Research Division, Pacific Grove, CA, USA

**Curtis H. Stumpf** Crystal Diagnostics Ltd., Rootstown, OH, USA

**Gabriela Valdivia** Department of Geography, University of North Carolina, Chapel Hill, NC, USA

**Carlos A. Valle** College of Biological and Environmental Sciences, Galapagos Academic Institute for the Arts and Sciences (GAIAS), Universidad San Francisco de Quito, Quito, Ecuador

**Byron Villacis** National Institute of Statistics and Census – INEC, Quito, Ecuador

**Julee Waldrop** School of Nursing, University of North Carolina at Chapel Hill, Chapel Hill, NC, USA

College of Nursing, University of Central Florida, Orlando, FL, USA

**Stephen J. Walsh** Department of Geography, Center for Galapagos Studies, Galapagos Science Center, University of North Carolina at Chapel Hill, Chapel Hill, NC, USA

**Wendy Wolford** Department of Development Sociology, Cornell University, Ithaca, NY, USA

# Abbreviations

| | |
|---|---|
| ABM | Agent-based model |
| ADAR | Airborne data acquisition and registration |
| ALI | Advanced land imager data |
| ANOVA | Analysis of variance |
| ASTER | Advanced spaceborne thermal emission and reflection |
| AVIRIS | Airborne visible/infrared imaging spectrometer |
| | |
| CAS | California Academy of Sciences |
| CDF | Charles Darwin Foundation |
| CDRS | Charles Darwin Research Station |
| CITES | International Convention for the Trade of Endangered Species |
| | |
| DEM | Digital elevation model |
| | |
| ENSO | El Niño southern oscillation |
| EOL | Encyclopedia of life |
| ETP | Eastern tropical pacific |
| | |
| FUNDAR | Foundation for Alternative Responsible Development |
| | |
| GAIAS | USFQ Galapagos Academic Institute for Arts and Sciences |
| GMR | Galapagos Marine Reserve |
| GNP | Galapagos National Park |
| GNPS | Galapagos National Park Service |
| GSC | Galapagos Science Center |
| | |
| INEC | Instituto Nacional de Estadistica y Censo |
| INEFAN | Institute of Forests and Natural Areas |
| INGALA | Galapagos National Institute |
| IPCC | Intergovernmental Panel on Climate Change |
| IUCN | International Union for the Conservation of Nature |
| IWC | International Whaling Commission |

JICA          Japan International Cooperation Agency

MODIS         Moderate resolution imaging spectroradiometer
MTMF          Mixture-tuned matched filter

NGO           Nongovernmental organization

OBIA          Object-based image analysis

SICGAL        Quarantine and inspection system for Galapagos
SST           Sea surface temperatures

UNC           University of North Carolina at Chapel Hill
UNESCO        United Nations Educational, Scientific and Cultural Organization
USAID         United States Agency for International Development
USFQ          Universidad San Francisco de Quito

WWF           World Wildlife Fund

# Chapter 1
# Science and Conservation in the Galapagos Islands

Carlos A. Valle

## Introduction

Darwin's Theory of Evolution, more than any other scientific theory, has changed our fundamental understanding of the living world and our own conception of human nature. It is no surprise that it has been considered the most revolutionary idea in modern science (Mayr 1982). The Galapagos Islands played a central role in the development of science, particularly for the field modernly known as evolutionary biology. It was mainly his observations on the species of this isolated group of islands in the Pacific that later led Darwin to doubt the chief concept of the time on the immutability of species and to reflect and develop his ideas about the origin of species and, subsequently, his theory of evolution by natural selection (Sulloway 1982). Ever since Darwin's memorable trip to the Galapagos Islands 175 years ago in September 1835, and the publication of his theory in 1859, a large number of naturalists and scientists have visited the Galapagos eager to test Darwin's observations and theory. The Swiss-American naturalist, Louis Agassiz of Harvard University, who visited the Galapagos in 1872; the German geologist Theodor Wolf, who visited twice, in 1875 and 1878; and George Bauer, who traveled there in 1891, are among the most renowned scientists in Darwin's time to visit the Galapagos (Larson 2001). During the twentieth century, two of the best known expeditions were the ones led by William Beebe, in 1923 and 1925, and the California Academy of Sciences expedition in 1905–1906 led by Rollo Beck (Larson 2001; Quiroga 2009).

Scientific research in the Galapagos was pioneered by Lack (1947), Bowman (1963, 1979) and Eibl-Eibesfeldt (1958). Thereafter, several dozen naturalists and scientists from around the world have visited and conducted research on Galapagos

C.A. Valle (✉)
College of Biological and Environmental Sciences, Galapagos Academic Institute
for the Arts and Sciences (GAIAS), Universidad San Francisco de Quito, Campus Cumbayá,
Diego de Robles y Vía Interoceánica, P.O. Box 17-12-841, Quito, Ecuador
e-mail: cvalle@usfq.edu.ec

S.J. Walsh and C.F. Mena (eds.), *Science and Conservation in the Galapagos Islands:*
*Frameworks & Perspectives*, Social and Ecological Interactions in the Galapagos Islands 1,
DOI 10.1007/978-1-4614-5794-7_1, © Springer Science+Business Media, LLC 2013

on a wide variety of subjects. Some came to fulfill an academic requirement like a dissertation, while a few others made the Galapagos their veritable home (e.g., see Quiroga 2009). Among the latter, there are celebrities like Peter and Rosemary Grant, now at Princeton University, who have visited the Galapagos for several months every single year for nearly 40 years since 1973 when their study of Darwin's finches began. Others include Tjitte de Vries, of the Universidad Católica del Ecuador, who pioneered research on the endemic Galapagos hawk (*Buteo galapa-goensis*), now, in collaboration with Patty Parker of the University of Saint Louis; Fritz Trillmich, of the University of Bielefeld, Germany, who researched the Galapagos fur seal (*Arctocephalus galapagoensis*) and the Galapagos sea lion (*Zalophus wollebaeki*); Andrew Laurie, whose work on the marine iguana (*Amblyrhynchus cristatus*) was continued and expanded by Martin Wikelski of the University of Konstanz; Tom Fritts, who researched the giant tortoises; Howard and Heidi Snell, who worked on land iguanas (*Conolophus subcristatus* and *C. pallidus*) and lava lizards (*Microlophus* spp.); and David Anderson, who carried out research on siblicide of the Nazca Booby (*Sula granti*) and the ecology of the Galapagos boobies (*Sula* spp.). Lately, a new generation of studies that takes advantage of modern molecular techniques has led to an explosion of phylogenetic and phylo-geographic studies that are clarifying our understanding of ecological and evolu-tionary processes, particularly, the processes and mechanisms of speciation, both in time and space.

The Galapagos Archipelago is one of the most studied places on Earth, even though much of its species' ecological and evolutionary processes still remain unknown. In the marine realm, very little is known about the intertidal, subtidal, and pelagic species and populations, as well as their communities' and ecosystems' ecol-ogy and evolutionary processes. The terrestrial biota has been better studied, but, even so, the knowledge base is strongly biased toward the most conspicuous and charismatic species, such as reptiles and some groups of birds. Among plants, only a few flowering plants have been relatively well studied, while very little is known about cryptic species such as cryptogamic plants (especially mosses and lichens).

Here, I briefly summarize some of the most relevant aspects of the natural history of the Galapagos by pointing out a few selected case studies that describe the most important research findings in different fields of the natural sciences studies con-ducted there. This mini review also addresses the conservation status and threats as well as the conservation achievements in the islands.

# Geological Research

Darwin can be considered the first geologist to observe and describe the geology of the Galapagos and to correctly affirm their geological youth and volcanic origin (emerging from the ocean) as oceanic islands that had not been in contact with con-tinental landmasses. Referring to the geology of the Galapagos, he once wrote, "We are led to believe that within a period, geologically recent, the unbroken ocean was here spread out." Modern geological research—aside from petrology and geochemical studies of basalts (e.g., Williams 1966; Williams and McBirney 1979; McBirney and Williams 1969; Swanson et al. 1974) and volcanology (Simkin and Howard 1970; Simkin 1984)—has focused mainly on the origin and age of the islands using several dating techniques including paleomagnetic polarities (geomagnetic reversals), geomorphology analyses, and radioactive potassium–argon dating (Cox 1971, 1983; Geist 1996). Modern radiometric work estimated that Española Island is about 3.3 million years old (Bailey 1976) and that the oldest islands of the archipelago range from about 2 to 3 million years old (Geist 2009).

Estimates of the ages of individual islands vary a lot, but the latest studies estimate the age of the oldest current extant islands (i.e., San Cristobal or Española) at about 3.5 million years and only about 60,000–300,000 years for Fernandina Island (Bailey 1976; Geist 1996). Once again, Darwin was correct when he inferred that the islands emerged from the unbroken ocean, as their origin is currently explained by hot spot and plate tectonic theories. Plate tectonics—the theory that continents move—was first proposed by de Candolle, a French biogeographer, who was originally discredited for his idea. However, it was formally proposed by Alfred Wegener in 1912 and was only gradually accepted after nearly 50 years, mostly due to paleomagnetic evidence. The hot spot theory, on the other hand, is a recent one and refers to a huge and extremely hot solid, but plastic, column of rock that probably rises from the deep mantle due to radioactive enrichment (and radioactive heat). The column (mantle plume) rises due to thermal buoyancy, melts near the surface owing to decompression, and breaks the Earth's crust from beneath, giving rise to a shield volcano like those found in the Galapagos, Hawaii, and other oceanic archipelagos.

The Galapagos settles over the Nazca plate that moves eastward to South America and runs under the South American continental plate along the so-called subduction zone. The islands travel from their center of origin (the Galapagos hot spot under Fernandina Island) at a variable rate between 2 and 7 cm per year. That has created a rough age gradient with a cluster of oldest islands in the east, a cluster of middle-aged islands in the center, and still another cluster of the youngest islands in the west of the archipelago (Simkin 1984). Although most of the Galapagos Islands owe their origin to the mantle hot spot, the origin of the two northernmost islands (Darwin and Wolf) and the three northwestern islands (Pinta, Marchena, and Genovesa) is likely related to the Galapagos ridge, the Nazca-Cocos plates spreading zone, which is located north of the Galapagos Islands (Geist 2009). Current marine geological exploration (Christie et al. 1992) has corroborated the hypotheses, first advanced by biologists (Wyles and Sarich 1983), that the older sunken Galapagos Islands may have been in existence for at least 10 million years. By now, most geologists and biologists working in the Galapagos readily accept that as the Nazca plate moves eastward and the islands travel away from the hot spot, they decrease in altitude—apparently owing to the cooling and contraction of the crust—and eventually subside (Geist 2009). Recent evidence about the genetic distance between the marine and land iguanas, and especially the new species of land iguana (Gentile et al. 2009), provides further support for the currently drawn Galapagos Islands and an older origin of the archipelago.

## Climate and Oceanographic Research

The Galapagos Islands are located where several marine currents converge, modifying what should be basically a tropical climate into a predominantly dry region during most of the year (Palmer and Pyle 1966). Such a geographical and oceanographic setting makes these islands attractive for oceanographers and climate scientists interested in the study of Earth's paleoclimate and for those attempting to understand and predict current climatic phenomena with a global impact, such as the El Niño–Southern Oscillation (ENSO) (e.g., Wyrtki 1975; Houvenaghel 1984; Cane 1983). The El Niño event, and its usual counterpart La Niña, affect the terrestrial and marine Galapagos biota in particular ways. During the conditions associated with El Niño (high sea surface temperature and a heavy rainy season), terrestrial organisms feast, breed freely, improve survival, and increase population sizes, while marine organisms fast and stop breeding, die, and decline in population. However, the reverse is true during the drop in sea surface temperature and drought conditions associated with La Niña (e.g., see Robinson and del Pino 1985).

Global-scale research has tremendously advanced our understanding of how the Earth's climate has changed over millions of years. Astronomic evidence and glacial–interglacial cycles, as well as local changes in landmass connectivity associated with plate tectonics, attest to the major climatic changes that the Earth has undergone on a global scale (e.g., Zachos et al. 2001; Ferodov et al. 2006). Thus, the opening and closing of the Isthmus of Panama had major consequences on the patterns of atmospheric and ocean circulation in the eastern Pacific (Cronin and Dowsett 1996). From these studies, it has been inferred how the climate of the Galapagos has changed over the last 10 million years from when the islands had a warmer and more humid tropical climate as compared to today.

Furthermore, studies conducted on the Galapagos Islands themselves have also improved our understanding of the climate both in these islands and at larger geographic scales, including the eastern Pacific. Most paleoclimate research on the Galapagos is based on analyses of sediment cores from a number of lakes but mainly at Junco Lake on San Cristobal Island, Lake Arcturus on Genovesa Island, and the lake on Bainbridge Rock islet near Santiago Island. Pollen stratigraphy and geochemical and mineralogical analyses of sediment cores and $C^{14}$ dating (Colinvaux 1968, 1972; Colinvaux and Schofield 1976a, b), carbon/nitrogen ratios, and isotopic hydrogen and oxygen analyses (Riedinger et al. 2002) have opened a window that has allowed us to picture the history of climatic changes in the Galapagos and the whole eastern Pacific Ocean. These studies provide evidence that during the last 50,000 years, the climate of the Galapagos has undergone profound changes in temperature and precipitation, in association with the Northern Hemisphere ice-age cycles. The islands were dry during glaciations and humid and rainy during interglacial periods, such as the present (Colinvaux 1984), and the frequency of El Niño events started increasing during the last 2,500 years, particularly in the last 1,000 years (Riedinger et al. 2002; Conroy et al. 2008). Research on lake sediments continues today with modern and improved devices and methods; research on El Niño and the climate dynamics of the Galapagos has also been tracked by studies of marine organisms including coral (Shen et al. 1992; Urban et al. 2000) and foraminifera (Lea et al. 2006). One of the major effects of climate change associated with glacial and interglacial cycles in the Galapagos relates to changes in sea levels. Sea level rises (interglacial) and falls (glacial) of over 100 m (Lambeck and Chappell 2001) are presumed to have affected levels of isolation, as well as the size and shape of the islands (Grant and Grant 2008); changes, that in turn, are expected to have affected evolutionary patterns and processes. In summary, global local-scale climate studies have shown that climate of the Galapagos has been highly dynamic and has changed dramatically over geological and evolutionary time scales.

## Research on Ecology and Evolutionary Biology

### *Research on Ecology*

Ecological research in the Galapagos is mainly of autoecological character and has focused on geographic distribution, demography, and the behavioral ecology of the most conspicuous species of terrestrial vertebrates (Bowman 1984; Clark 1984; Eibl-Eibesfeldt 1984a, b; de Vries 1984; Grant 1984, 1999; Grant and Grant 2008; Harris 1984; Trillmich 1984; de Roy 2009) and a few species of vascular plants (Eliasson 1984; Porter 1984; Tye 2007, 2008). Even so, several gaps remain regarding the ecology of these relatively well-studied groups, and even less is known about most species of terrestrial invertebrates and most vascular and nonvascular plants, as well as most marine organisms. Furthermore, except for a few marine studies (Withman and Smith 2003; Edgar et al. 2004) and a handful of terrestrial organisms (Abbott and Abbott 1978; Schluter 1986; Schluter and Grant 1984; Schluter et al. 1985; Grant and Grant 2006), ecological research in the Galapagos has severely neglected community ecology at the ecosystem level, with almost a complete lack of in-depth studies about interspecific interaction at the community level and its role on the structuring of natural communities. Thus, we know very little about ecological processes such as pollination, seed dispersal, symbiotic interactions (e.g., parasitism, commensalism), competition, predation, decomposition, predator-/herbivore-mediated coexistence, and the occurrence and role of ecological guilds.

### *Research on Evolutionary Biology*

A volcanic, highly isolated, oceanic archipelago like the Galapagos that was never connected to the mainland is expected to have been devoid of terrestrial life when it emerged from the ocean. The terrestrial life now inhabiting these islands, as well as those organisms now extinct, got there from somewhere else and became rapidly and effectively isolated from their parental populations. These two aspects of colonization and further isolation have profoundly shaped the ecological and evolutionary history of the Galapagos organisms through a three-step process involving (1) organisms' arrival, (2) establishment (colonization) in their new habitat, and (3) in situ evolution (i.e., adaptive and nonadaptive genetic and phenotypic divergence, speciation, evolutionary radiation).

## Arrival of Organisms to the Archipelago

Arrival requires transportation to the islands from somewhere else, and 1,000 km of ocean represents a major ecological barrier for a vast number of organisms. A key factor for dispersal and arrival was the organisms' vagility (dispersal ability), which is dependent upon their intrinsic dispersal abilities (i.e., production of small propagules that disperse easily or are easy to transport by winds or by animals) and their ability to survive a long journey through the ocean. Organisms reached the islands by three means of dispersal: through the ocean, either by floating and drifting with the ocean currents (passive dispersal) or swimming (active dispersal); through the air, taking advantage of the trade winds blowing from southeast and northwest; or by attaching themselves to other organisms (e.g., birds). Many organisms, including several types of invertebrates (snails, arthropods) and vertebrates (lizards, snakes, and even poor-flying birds like rails), may have reached the islands on those masses of vegetation that usually get to the ocean when continental rivers overflow during the rainy season and then drift into the open ocean. Duncan Porter (1976) inferred the mean of transportation for the 378 indigenous ancestral taxa that colonized the Galapagos and suggests that birds may have been the vector for about 60% (of these, 64% in the digestive tract, 21% attached to feathers, and 15% attached to mud on the legs), while 32% were transported by the wind and 9% by the sea.

## Establishment of Organisms in the Archipelago

Getting to the Galapagos was only part of the process; colonization required the establishment of a viable population. It can be inferred that many successful arrivals did not lead to a successful colonization due to a number of factors, including (1) the ecological successional stage of the island, (2) reproductive viability, and (3) demographic (population) viability. Many organisms may have been prevented from settling after not finding a suitable habitat upon arrival. For example, as we can see today on a barren lava flow, only a few pioneer species (e.g., lichens and few other plants) would be expected to have established themselves first, while the more habitat-demanding species would have had to wait for conditions that were more suitable. For organisms with strict sexual reproduction (most animals and monoecious plants), an unavoidable condition for establishment was arriving with a mate, or in a small flock with individuals of both sexes (as may be the case of the birds), or a fertilized female. Populations that are too small and isolated, such as those that just have colonized a remote archipelago from one single flock, a fertilized female, or a single asexual individual, run a high risk of extinction due to stochastic factors (including genetic, demographic, and environmental), as well as the occurrence of natural catastrophes (Lande 1980; Lande and Barrowclough 1987).


## Evolution in the Galapagos Archipelago

Perhaps the most prevalent biological feature of isolated oceanic archipelagos is the high level of endemism among every native taxonomic group. Insular endemics evolve from a mainland-colonizing ancestor either through *linear evolution*, when a colonizing ancestor transforms into a new insular species, or through an *evolutionary radiation*, when either an ancestral colonizer or an insular endemic splits further into several new species.

In the Galapagos Islands, most species that colonized the archipelago have evolved at least into single endemic taxa including subspecies, species, and genera (see Baert 2000; Peck 1996; Tye et al. 2002). Endemics at the species level included mammals (~88%), birds (52%), reptiles (100%), fishes and algae (~20–30%) insects (47%), other terrestrial invertebrates (53%) and vascular plants (32%), bryophytes (mosses, liverworts; ~10%), lichens (~7%), and pteridophytes (ferns; ~4%). Examples of a single colonizing ancestor evolving into a new species (an insular endemic) abound, including the two species of sea lions, the Galapagos penguin (*Spheniscus mendiculus*), the flightless cormorant (*Phalacrocorax harrisi*), and several others. A lower number of colonizers evolved even further and underwent evolutionary radiation when a single colonizing species evolved into several new endemic species. The best examples of evolutionary radiation, most of them derived from a single colonizing event (i.e., monophyletic groups), include land snails of the genus *Bulimulus* (71 species, all endemic, perhaps the most spectacular example of adaptive radiation), Darwin's finches (15 species), and giant tortoises (originally about 15 species). Among plants, the genus *Scalesia* (Asteraceae; 15 species, 19 taxa including subspecies and varieties) and *Alternanthera* (Amaranthaceae; 14 species and 20 taxa), *Opuntia* (Cactaceae; six species and 14 varieties, from two independent colonizations).

The geological youth of the Galapagos Archipelago, together with the high level of endemism among all taxonomic groups that colonized it, implies that evolution in the Galapagos has generally proceeded rapidly particularly among those species that have evolved into different species through the process of linear evolution. The rate of evolution (speciation) has been even more dramatic among those groups that have radiated into many species from a single ancestor that colonized the islands.

The process of colonization and evolution of all organisms in the Galapagos fits nicely into the *founder effect speciation* model first suggested by Huxley (1938) and formally proposed by Mayr (1954). That is, a new species arises when a new population is founded in a remote isolated place, usually by a small group on immigrants. Under such a model, theoretically, there are a number of ecological and genetic factors and processes driving rapid evolution in an isolated archipelago like the Galapagos. These factors and processes include *founder effect, genetic drift, divergent selection, interrupted or restricted gene flow, multiple isolation,* and *ecological opportunity.*

*Founder Effect*: The role of the founder effect in immediately driving genetic divergence and subsequent phenotypic divergence between a founding (island) and a parental (continental) population is expected to be high; few individuals will bring, at best, only a small fraction of the whole genetic variation from a large parental population simply due to random genetic sampling error. Furthermore, a very small number of founding individuals may not necessarily be a representative sample of the whole parental population (e.g., Huxley 1938). A genetically nonrepresentative sample means that the island's founding population will start with a biased sample of alleles relative to the continental population. All of this suggests that a new insular founding population is likely to diverge from its continental parental population from the very beginning. Founder effects are also expected to lead to inbreeding, due to the small number of colonists. Inbreeding is predicted to lead to increased homozygosis in the population, which would potentially lead to inbreeding depression, thus increasing the probability that the founding population could become extinct. Only those populations that managed to go through such a demographic bottleneck and survive were the ones that become established (colonized) in the Galapagos. Such populations, at least on theoretical grounds, may be able to survive, evolve, and remain well adapted through a coadapted set of genes; for them, outbreeding could be deleterious in the sense that it would break the coadapted set of genes thus reducing or wiping out their adaptive value. Over time, however, most populations are expected to increase in number and in genetic variation.

Owing to the development of molecular genetics, studies only recently started to address this issue, as well as that of the role of genetic drift in the process of evolution in the Galapagos. The few studies addressing this issue seem to indicate that some species may in fact have started with extremely low numbers (i.e., a founder effect genetic signature has been found). On Daphne Major Island, Grant and Grant (2008) observed and neatly tracked a founder event of a population of the large ground finch (*G. magnirostris*) that had colonized the islet during the last 25 years.

*Genetic Drift*: Genetic drift, a random fluctuation of allele frequencies leading to an eventual random loss/fixation of alleles in a population, is expected to play an important role in the evolution of small and isolated populations (Wright 1931, 1932). The Galapagos' recently founded populations that have experienced a demographic bottleneck due to a founder effect would constitute an ideal scenario for drift to occur. Thus, colonizing alleles would become lost/fixed at a faster rate, leading to rapid further genetic divergence between the island and the continental population. Such a stochastic evolutionary process would result in nonadaptive evolution, which is theoretically expected to be a common pattern at the onset of colonization in a place like the Galapagos. Genetic erosion or loss of alleles (a consequence of drift) and inbreeding (a consequence of a founder effect) are expected to lead to increased homozygosis in the population, perhaps contributing in turn to inbreeding depression and increasing chance of extinction.

*Divergent Selection*: Isolated populations are expected to diverge rapidly as a result of divergent natural selection for local adaptation. Selection pressures usually vary geographically and doubtless are expected to differ greatly between the continent and a new colonized insular habitat, as well as from island to island. Selection, acting upon a biased and impoverished genetic variation (due to founder effect and genetic drift), as well as upon new alleles (i.e., originated by independent mutation) and new genotypes (due to recombination), will rapidly or gradually lead to genetic and phenotypic divergence between insular and continental populations. Such divergence is expected to happen rapidly among small and isolated populations or rather gradually among much larger populations.

*Interrupted or Restricted Gene Flow*: Gene flow is a homogenizing evolutionary force that prevents conspecific populations from diverging due to local adaptation. Therefore, if gene flow became effectively interrupted or severely restricted due to geographical isolation, genetic divergence between source (continental) and founder (island) population would be rapid. Recent molecular genetics and ecological inferences suggest that, in this sense, the almost 1,000 km of open ocean have acted as an effective ecological barrier preventing gene flow for most of the terrestrial organisms that colonized the Galapagos. Founder effect, genetic drift, and divergent selection, aided with lack of gene follow, will result in genetic and phenotypic divergence between conspecific populations, eventually leading to speciation and the origin of endemic taxa (species or subspecies).

*Multiple Isolation*: Archipelagos, as opposed to single islands, present opportunities for multiple and repeated events of dispersal and colonization followed by further isolation (an *archipelago effect*). This means that within archipelagos, a natural evolutionary experiment of colonization of one island from the mainland, as described above, is replicated every time that a species already present in the archipelago disperses and colonizes a new island. Such a combination of ecological and genetic factors and processes repeated several times on different islands would be another ingredient for rapid evolutionary diversification of a lineage into several endemic taxa (genera, species, subspecies). The Galapagos, where single colonization events have led to evolutionary radiation in a number of endemic organisms (e.g., Darwin's finches, bulimulid land snails, *Scalesia* plants, and *Opuntia* cacti), fits these theoretical expectations very closely.

*Ecological Opportunity*: Oceanic archipelagos that were originally devoid of terrestrial life, like the Galapagos, will only gradually fill with species that arrive and establish on the islands. Insular ecological communities, therefore, are expected to be less packed than continental ones, especially at the early and middle stages of their ecological succession. These more relaxed ecological communities will have a number of *empty* ecological niches that will not only favor continuous colonization of new species (McArthur and Wilson 1967) but will also allow already established species the opportunity to *explore* and, in some cases, eventually shift into a new ecological niche. Species that show large genetic variation for morphological and behavioral adaptive traits, such as Darwin's finches (e.g., Grant and Grant 1989, 2008), will be more likely to undertake such an evolutionary path and thus radiate into several new species, each one adapted to a particular ecological niche (Lack 1947; Grant 1999; Grant and Grant 2008).

On theoretical grounds, in isolated archipelagos like the Galapagos, genetic drift has the potential to be a relevant factor in evolution, particularly following a founder event. Thereafter, because the population increases in size or becomes less isolated within the archipelago, the effect of drift as an evolutionary force is expected to decline, while natural selection is likely to catch up and become the chief mechanism of further divergence. This does not imply that natural selection may not be an important evolutionary force from the very beginning following a founder event.

In the past, evolutionary ecologists usually assumed that all or most speciation resulted from Darwinian (adaptive) evolution and quickly accepted most evolutionary radiations as putative examples of adaptive radiation. Most modern-day evolutionary biologists would accept that a combination of stochastic processes (i.e., drift and founder effect) and adaptive (Darwinian selection) mechanisms are usually involved in evolution. The relative importance of natural selection and stochastic processes, however, is still a matter of debate on both theoretical and empirical grounds. Whether evolutionary radiations are mainly adaptive or the product of stochastic process still deserves in-depth research in the Galapagos and elsewhere.

Some divergent traits among closely related species [e.g., morphological and behavioral traits such as song and beak size and shape in Darwin finches (Lack 1947; Bowman 1979; Grant 1999; Grant and Grant 2008)] show a clear adaptive function and evolutionary divergence. Speciation among those lineages can safely be attributed to the role of selection although that does not imply that the role of drift at some point on their evolutionary history should be disregarded. The adaptive value of other traits is less obvious since there is not an apparent function for survival and their evolution may be either the result of sexual selection or genetic drift. Geographical and interspecific divergence at the molecular level (i.e., DNA sequences) is regarded to be mainly the result of genetic drift, and, although the issue is still controversial, new molecular techniques and statistical methods have started to reveal evidence for selection at this level.

Most biological research on the Galapagos has been centered on the evolutionary ecology of vertebrates and a few invertebrates and plants. The evolutionary biology of marine organisms largely remains an unexplored field. Modern molecular genetic techniques have prompted a new generation of evolutionary studies in the Galapagos, including fields such as population genetics, phylogenetics, phylogeography, and evolutionary developmental biology. These new and recent studies are shedding light on the patterns of genetic variation within and between populations, information that is important for conservation. These studies are also clarifying the taxonomic status of many different taxa, such as the taxonomic position of each species (previously considered as subspecies) and the finding of cryptic taxa of giant tortoises. Just to list a few, the taxonomic positions of species whose Galapagos populations are now accepted as full new and endemic species include the Nazca booby (*Sula granti*), the Galapagos petrel (*Pterodroma phaeopygia*), the Galapagos shearwater (*Puffinus subalaris*), and the green heron (*Butorides striatus*). More than that, these studies are allowing the exploration of the evolutionary history of Galapagos organisms, both in space and time. These studies have begun to clarify more accurately several issues that were previously difficult to track, such as the ancestral species and geographical origins of the different native species, especially of those that have radiated. We have still a long way to go, but thanks to molecular genetic studies, at least we have gained a better understanding of the historical evolution of Darwin's finches (Petren et al. 1999, 2005; Sato et al. 2001a, b), Galapagos mockingbirds (Arbogast et al. 2006; Hoeck et al. 2010), Galapagos hawk (Bollmer et al. 2005, 2006, 2007; Whiteman et al. 2007; Hull et al. 2008; Parker 2009a, b), Galapagos cormorant (Kennedy et al. 2009; Duffie et al. 2009), giant tortoises (Caccone et al. 1999, 2002), land and marine iguanas (Wyles and Sarich 1983; Rassmann 1997; Gentile et al. 2009), lava lizards (Lopez et al. 1992; Jordan et al. 2002; Kizirian et al. 2004; Jordan and Snell 2008), bulimulid land snails (Parent and Crespi 2006; Parent et al. 2008), and a few others (Sequeira et al. 2000; Schmitz et al. 2007).

## Biological Conservation

Oceanic archipelagos and island ecosystems, in general, are highly vulnerable to disturbance, especially to invasion by exotic (introduced) organisms (Crawley 1987). Such a high level of vulnerability is likely explained by a history of evolution in isolation from the mainland (i.e., insular organisms evolved free from major and diversified competition, predation, and disease) that has resulted in their ecosystems' low resistance and low resilience (Carlquist 1965; Connell and Sousa 1983). Other causes for their ecosystems' fragility may be explained by a pattern of low species diversity, low complexity and demographic factors, including small population size and restricted distribution range for a large number of island species and the existence of vacant niches (Connell and Sousa 1983; Herbold and Moyle 1986; Mace and Lande 1991).

The Galapagos Archipelago fully fits these generalizations of being ecologically fragile. The islands' ecosystems are species-poor and simple (Snell et al. 2002); however, relative to its size, the Galapagos' contribution to global biodiversity is high, due to a high endemism among all taxa. Also, the Galapagos biota evolved largely in isolation from mainland South America with very low presumed rates of natural arrival and colonization of species over their geological history. On the demographic side, a large number of native species of plants and animals have an extremely small population size and/or distributions restricted to a single island or even to only a small area within an island. The population size of several of these species, as well as their distribution areas, has been further reduced due to current threats.

## Conservation Threats

The Galapagos' ecosystems are under pressure on two fronts: terrestrial and marine, each one having its own peculiarities.

## Terrestrial Ecosystems

The main threat to terrestrial ecosystems is the large and diversified number of exotic organisms (plant, animals, bacterial, and viral diseases), many of which have become invasive. Most of the introduced organisms are competitors, predators, parasites for native species, vectors, or reservoirs of diseases that later spread to organisms. The ecological impact of exotic organisms is worsened, because native species have not usually evolved immunological defenses against recently introduced diseases or developed behaviors and other life-history strategies to counteract the effects of exotic predators and competitors. The problems of exotic species in the Galapagos are further worsened because, while roads are the main means of dispersing exotic plants and animals, current development trends aim to build new roads across each inhabited island.

Exotic organisms have been introduced to the islands voluntarily or involuntarily by humans for 400 years, since pirates started using the islands regularly. The rate of introductions has worsened during last two decades, mainly due to a dramatic increase in the resident human population and the increased rate of transportation both within the archipelago and between the archipelago and the continent (Tye et al. 2008; Causton and Sevilla 2008). Contrasting with 112 introduced species recorded by 1900, the Charles Darwin Research Station and the Galapagos National Park confirmed the establishment of 748 species of vascular plants (cf. ~500 native), 543 invertebrates (cf. ~3,000 native), mostly insects (490 introduced cf. 1,555 native), and 30 vertebrates (Tapia et al. 2000; Roque-Albelo 2008; Causton and Sevilla 2008; Tye 2007, 2008; Tye et al. 2008) by 2007. Most remain in the place they were introduced on the five human-inhabited islands and within the colonized zones of these islands, but they are gradually spreading over much of the archipelago. By now, none of the 19 larger islands of the archipelago are free from introduced organisms.

The worst introduced species are the most invasive ones that rapidly spread from their center on introduction. Among these are hill blackberry (*Rubus niveus*) and guava (*Psidium guajava*), which are invading extensive areas of the humid zone on San Cristobal, Santa Cruz, Floreana, and Isabela Islands, and quinine (*Cinchona pubescens*), which is widespread on Santa Cruz (Tye et al. 2008). Their negative impact on animals is widely recognized (Causton and Sevilla 2008; Jiménez-Uzcátegui et al. 2008a, b). Goats (*Capra hircus*) compete for food with giant tortoises, causing the decline of a number of rare plants and provoking the decline of otherwise rather common plants, such as the cactus trees (*Opuntia* sp.) in islands like Santiago which, in turn, endangers the cactus finch (*Geospiza scandens*), which is highly dependent on the cactus for survival. Rats (black *Rattus rattus*; Norwegian, *Rattus norvegicus*), dogs (*Canis familiaris*), cats (*Felis catus*), and pigs (*Sus scrofa*) are very active and voracious predators, attacking insects (e.g., native beetles), land and marine birds and even nests and hatchlings of birds, and all reptiles, including giant tortoise hatchlings, lava lizards, land and marine iguanas, snakes, and sea turtles. The recently introduced ani (*Crotophaga ani*) spread through the archipelago during the El Niño of 1982–1983 (Valle, personal observations), likely causing the population to decline and perhaps even causing future extinctions of several land birds (e.g., the vermilion flycatcher), although this issue needs further investigation.

Introduced invertebrates seem to be the worst among the animals, because of their rapid spread and their apparent impact on the invertebrate and vertebrate community (Causton and Sevilla 2008; Roque-Albelo 2008; Fessl et al. 2006). Of primary concerns are two fire ants (*Wasmannia auropunctata* and *Solenopsis geminata*) and two wasps (*Brachygastra lechiguana, Polistes versicolor*). The cottony cushion scale (*Icerya purchasi*) spread rapidly along the coastal zone in Santa Cruz and caused the death of mangroves and several other plant species. The fly (*Philornis downsi*) infests birds in their nests. The potential for further introduction is remarkably high due to the high volume of organic products (mostly fruits and vegetables) brought to the islands every week as food supply for the local population and tourists.

Diseases pose a major threat for oceanic fauna, as exemplified in Hawaii, where more than one-half of native birds became extinct from avian malaria (Van Riper et al. 1988). In the Galapagos, introduced vertebrates, particularly birds and mammals, are known vectors or reservoirs of several viral, bacterial, and protozoan diseases (Vargas and Snell 1997; Miller et al. 2001; Wikelski et al. 2004). A viral disease carried by black rats was the most likely reason for the extinction of most Galapagos native rice rats. Viral and protozoan diseases are the subject of an investigation started by the Charles Darwin Foundation (Vargas and Snell 1997) and now continued by Patricia Parker's team (Gottdenker et al. 2005; Whiteman et al. 2005; Parker et al. 2006, 2009a, b; Duffie et al. 2008; Santiago-Alarcon et al. 2008; Levin et al. 2009).

Habitat degradation and loss is probably the second most important threat for terrestrial biodiversity in the Galapagos and the most likely direct cause of recent extinctions among some native endemic invertebrates and vertebrates (especially land birds). The humid zones of the four major islands (which also happen to be the islands inhabited by humans) are probably the most important habitats for most exclusive terrestrial Galapagos organisms, both vertebrates and invertebrates. The vegetation zone that remained largely unaltered up to the 1970s has now almost been completely replaced by introduced pastures and other invasive exotic plant species, especially on Santa Cruz and San Cristobal and to a lesser extent on Isabela and Floreana where large areas have been invaded also. The removal and alteration of natural habitats in these areas have severe implications for the conservation of a large number of native species. This alone, or in combination with other factors, may be the cause of the extinction of two species of endemic snails and the virtual disappearance of the vermilion flycatcher on San Cristobal and its apparent decline in Santa Cruz.

## Marine Ecosystems

The Galapagos marine ecosystems remained largely pristine until very recently, in spite of heavy exploitation during the nineteenth century by whalers, fur sea lion hunters and, since the early 1900s, tuna fishers (Larson 2001). Besides the direct and severe effect on the exploited populations (i.e., whales and the Galapagos fur sea lion *Arctocephalus galapagoensis*), the ecosystem as a whole was not largely impacted, and apparently, at least for fur seals, these exploited populations recovered in full (Trillmich, personal communication). Recent research suggests that the Galapagos sea lion (*Zalophus wollebaeki*) population may be declining; however, preliminary results await further confirmation since methodologies between past and recent censuses differed substantially (Bustamante et al. 2002). Marine predators and grazers, both among vertebrates and invertebrates are, however, affected negatively by recurrent El Niños. The most severe effects of El Niño in the Galapagos were recorded during 1982–1983. In that year, most sea birds ceased reproduction, experienced an unprecedented high mortality, and some species including the Galapagos penguin (*Spheniscus mendiculus*) and the Galapagos cormorant (*Phalacrocorax harrisi*) suffered the most severe population declines ever recorded (Valle 1985; Valle and Coulter 1987; Valle et al. 1987). A similar pattern, although without the dramatic decline of penguins and cormorants, was also recorded for the Galapagos fur seal and the Galapagos sea lion (Limberger 1985; Trillmich and Limberger 1985; Trillmich 1985), marine iguana (*Amblyrhynchus cristatus*) (Laurie 1985), fishes (Grove 1985), corals that almost disappeared (Glynn 1986, 1994), and several other organisms (see Robinson and del Pino 1985). Another exceptionally strong El Niño took place in 1997–1998 that also led to interruptions and mortality and population declines of a large number of marine organisms (Vargas et al. 2006). Marine biologists suggest that the effects of the El Niño and anthropogenic factors combined to threaten several marine organisms, which in turn are severely altering the community composition of Galapagos marine ecosystems (Branch et al. 2002). Although it is still debatable, some researchers think that the frequency and intensity of El Niños may be strengthening due to anthropogenic factors at the global scale.

The main current threat to marine ecosystems is overfishing. The targets of traditional small-scale fisheries—the grouper bacalao (*Mycteroperca olfax*) and three species of lobster (*Panulirus penicillatus*, *P. gracilis*, *Scyllarides astori*)—are already overexploited. Sea cucumber (*Stichopus fuscus*) fishing began as a high income opportunity in the early 1990s, but became overexploited in less than a decade, as some predicted (Valle 1994). By now it has become economically unprofitable. Although baseline data are still scanty, circumstantial evidence strongly suggests that several marine invertebrates of the intertidal and subtidal zones (e.g., snails, crabs, chitons, octopi) around local communities are heavily impacted by on-foot fishing. A preliminary assessment failed to detect any impact from tourism on the marine visitor sites (Bustamante et al. 2002). However, there is growing concern that marine subtidal zones at visitor sites may become affected especially by the dramatic increase in tourism over the last 10 years. All ecosystems, but particularly the marine ones, face the potential impact from accidental fuel spills from the nearly 100 tourist boats, half-dozen cargo boats, and tankers that provide fuel to the islands for a remarkably fast-growing number of automobiles. The most striking fuel spill accident was that of the Jessica tanker in January 2001, when the tanker runs aground on Wreck Bay on San Cristobal and more than 240,000 gallons of fuel spilled into the ocean.

## Conservation Achievements

Much effort has been devoted to the conservation of the Galapagos by both Ecuador and the world. Ecuador pioneered conservation efforts in South America when it declared a number of island protected areas in the 1930s and, subsequently, when it declared all terrestrial land that was not yet colonized in the Galapagos by 1959 (97.3% of the land area) as a national park. Another conservation landmark was an agreement signed by Ecuador and the Charles Darwin Foundation that allowed the creation of the Charles Darwin Research Station, which began its operation in 1964. Since 1964, both the Galapagos National Park and the Charles Darwin Foundation (through its operative arm, the Charles Darwin Research Station) have exerted tremendous effort to fulfill their mission of conserving the Galapagos Islands. Their main conservation efforts focused on (1) the protection and restoration of native endangered species and habitats; (2) the control and eradication of exotic species, focusing particularly on the most invasive flora and fauna; and (3) the environmental education of a rapidly growing resident population.

Achievements in protecting and restoring native endangered species has mainly taken place via the captive breeding program for land iguanas and giant tortoises. Their subsequent repatriation into their natural habitat qualifies as an unprecedented conservation success. The most celebrated case is that of the Espanola giant tortoise (Milinkovitch et al. 2004) that was at the brink of extinction in 1964 with only 14 individuals in the wild (12 females and two males) and a third male at the San Diego Zoo, all of whom were brought to the Darwin Station for breeding in captivity; by 2010, nearly 2,000 young tortoises had been repatriated and a few of the first repatriated young already had started breeding in the wild (Milinkovitch et al. 2004). There is, however, still room for concern about the long-term viability of this population, due both to the small number of parents, the consequent low levels of heterozygosis, and a small effective population size.

Eradication of feral introduced vertebrates, including goats (*Capra hircus*), pigs (*Sus scrofa*), dogs (*Canis familiaris*), cats (*Felis catus*), donkeys (*Equus asinus*), and pigeons (*Columbia livia*) has been successful on a number of small islands and at particular locales on the larger islands (Tapia et al. 2000; Jiménez-Uzcátegui et al. 2008a, b). Another notable conservation achievement was the *Isabela Project*, a well-planned intensive eradication project that led to the eradication of goats on northern Isabela Island, a vast area of difficult terrain, and the virtual eradication of goats on Pinta Island and goats and pigs on Santiago Island (Campbell et al. 2004). The local control of introduced species, particularly mammals, has proved effective in particular locations at decreasing predation and increasing in situ reproductive success and population numbers for a large number of native species including the Galapagos Petrel (*Pterodroma phaeopygia*) on Floreana and Santa Cruz Islands, land iguanas on Santa Cruz and Isabela Islands, and giant tortoises on several islands. Along the same lines, the first time biological control systems was applied in the Galapagos to control the introduced cotton cushion scale (*Icerya purchasi*) that started devastating native vegetation with the introduction and controlled release of the ladybug beetle (*Rodolia cardinalis*).

The Galapagos National Park and the Charles Darwin Research Station have several eradication projects underway aiming to eradicate or control at least the most invasive introduced plants, vertebrates, and invertebrates. Enormous further investment is needed to succeed, especially with the need to fight not only invasive species already on the islands but also new ones that are being introduced every year, due to an ineffective and poorly implemented quarantine control.

On the marine side, the creation in 1998 of the Galapagos Marine Reserve (GMR), one of the world's largest marine reserves backed by an organic special law, was, without a doubt, the greatest marine conservation achievement. The GMR which embraces 140,000 km² of marine waters, both within and around the islands, is a multiple-use reserve (protection, small-scale extractive and non-extractive activities are allowed) that excludes industrial fishing and confers a high level of protection to marine ecosystems and species within 40 miles surrounding the Galapagos. The GMR is under the administration of the Galapagos National Park directorate but has several bodies that allow a process of participatory management where local stakeholders are part of the decision-making process. To implement the management of the reserve, a management plan was developed and an on-the-ground zoning system of the coastal waters was put in place and is being implemented.

Besides a good level of scientific knowledge relevant to conservation, the Galapagos has the legal framework and elements to achieve its conservation goals in the long run. In Ecuador, Galapagos is the only province with an organic special law that declares and promotes conservation and sustainable development as its fundamental principles. However, due to a general lack of law enforcement due to limited resources and limited willingness, the long-term conservation of the Galapagos cannot yet be guaranteed. The islands were included in the 2007 list of World Heritage sites in peril. There is an urgent need for an Ecuadorean state policy for the conservation of the islands. Although there is some disagreement about how much of the Galapagos native biota is still in place, nobody doubts that the Galapagos remains one of the most pristine places on Earth, something that humanity cannot afford to lose.

# References

Abbott I, Abbott LK (1978) Multivariate study of morphological variation in Galapagos and Ecuadorian mockingbirds. Condor 80:302–308

Arbogast BS, Drovetski SV, Curry LR, Boag PT, Seutin G, Grant PR, Grant BR, Anderson DJ (2006) The origin and diversification of Galapagos mockingbirds. Evolution 60:370–382

Baert L (2000) Invertebrate research overview: 1. Terrestrial arthropods. In: Sitwell N, Baert L, Coppois G (eds) Science for conservation in Galapagos. Bulletin van het Koninklijk Belgisch Instituut voor Natuurwetenschappen, Suplement 70:23–25

Bailey K (1976) Potassium–argon ages from the Galapagos Islands. Science 192:465–466

Bollmer JL, Whiteman NK, Cannon MD, Bednarz JC, de Vries T, Parker PG (2005) Population genetics of the Galapagos Hawk (*Buteo galapagoensis*): genetic monomorphism within isolated populations. Auk 122:1210–1224

Bollmer JL, Kimball RT, Whiteman NK, Sarasola JH, Parker PG (2006) Phylogeography of the Galapagos hawk (*Buteo galapagoensis*): a recent arrival to the Galapagos Islands. Mol Phylogenet Evol 39:237–247

Bollmer JL, Vargas FH, Parker PG (2007) Low MHC variation in the endangered Galapagos penguin (*Spheniscus mendiculus*). Immunogenetics 59:593–602
Bowman RI (1963) Evolutionary patterns in Darwin's finches. Occas Pap Calif Acad Sci 44:107–140
Bowman RI (1979) Adaptive morphology of song dialects in Darwin's finches. J Ornithol 120:353–389
Bowman RI (1984) Contributions to science from the Galapagos. In: Perry R (ed) Key environments: Galapagos. Pergamon Press, Oxford, pp 277–311
Branch GM, Whitman JD, Bensted-Smith R, Bustamante RH, Wellington GM, Smith F, Edgar GJ (2002) Conservation criteria for the marine biome. In: Benson-Smith R (ed) A biodiversity vision for the Galapagos Islands. Charles Darwin Foundation and World Wildlife Fund, Puerto Ayora, Galapagos, pp 80–95
Bustamante RH, Branch GM, Bensted-Smith R, Edgar GJ (2002) In: Benson-Smith R (ed) A biodiversity vision for the Galapagos Islands. Charles Darwin Foundation and World Wildlife Fund, Puerto Ayora, Galapagos, pp 80–95
Caccone A, Gibbs JP, Ketmaier V, Suatoni E, Powell JR (1999) Origin and evolutionary relationships of giant Galapagos tortoises. Proc Natl Acad Sci USA 96:13223–13228
Caccone A, Gentile G, Gibbs JP, Fritts TH, Snell HL, Betts J, Powell JR (2002) Phylogeography and history of giant Galapagos tortoises. Evolution 56:2052–2066
Campbell K, Donlan CJ, Cruz F, Carrion V (2004) Eradication of feral goats *Capra hircus* from Pinta Island, Galapagos, Ecuador. Oryx 38:1–6
Cane MA (1983) Oceanographic events during El Niño. Science 222:1189–1194
Carlquist S (1965) Island life. Natural History Press, Garden City, New York
Causton C, Sevilla C (2008) Latest records of introduced invertebrates in Galapagos and measures to control them. Galapagos report 2006–2007. CDF, GNP and INGALA, Puerto Ayora, Galapagos, Ecuador, pp 142–145
Christie DM, Duncan RA, McBirney AR, Richards MA, White WM, Harpp KS, Fox CG (1992) Drowned islands downstream from the Galapagos hotspot imply extended speciation times Nature 355:246–248
Clark D (1984) Native land mammals. In: Perry R (ed) Key environments: Galapagos. Pergamon Press, Oxford, pp 225–231
Colinvaux PA (1968) Reconnaissance and chemistry of the lakes and bogs of the Galapagos Islands. Nature 219:590–594
Colinvaux PA (1972) Climate and the Galapagos Islands. Nature 240:17–20
Colinvaux PA (1984) The Galapagos climate: present and past. In: Perry R (ed) Key environments: Galapagos. Pergamon Press, Oxford, pp 55–69
Colinvaux PA, Schofield EK (1976a) Historical ecology in the Galapagos Islands. I. A Holocene pollen record from Isla San Cristobal. J Ecol 64:989–1012
Colinvaux PA, Schofield EK (1976b) Historical ecology in the Galapagos Islands. II. A Holocene spore record from Isla San Cristobal. J Ecol 64:1013–1026
Connell JH, Sousa WP (1983) On the evidence needed to judge ecological stability or persistence. Am Nat 121:789–824
Conroy JL, Overpeck JT, Cole JE, Shanahan TM, Steinitz-Kannan M (2008) Holocene changes in eastern tropical Pacific climate inferred from a Galapagos lake sediment record. Quaternary Sci Rev 27:1166–1180
Cox A (1971) Paleomagnetism of San Cristobal Island, Galapagos. Earth Planet Sci Lett 11:152–160
Cox A (1983) Ages of the Galapagos Islands. In: Bowman RI, Berson M, Leviton AE (eds) Patterns of evolution in Galapagos organisms. American Association for the Advancement of Science, Pacific Division, San Francisco, CA, pp 123–155
Crawley MJ (1987) What makes a community invasible? In: Gray AJ, Crawley MJ, Edwards PJ (eds) Colonization, succession and stability. Blackwell Scientific, Oxford, pp 429–453
Cronin TM, Dowsett HJ (1996) Biotic and oceanographic response to the Pliocene closing of the Central American Isthmus. In: Jackson JBC, Budd AF, Coates AG (eds) Evolution and environment in tropical America. University of Chicago Press, Chicago, pp 76–104

De Roy T (ed) (2009) Galapagos: preserving Darwin's legacy. David Bateman Ltd., New Zealand
De Vries T (1984) The giant tortoises: a natural history disturbed by man. In: Perry R (ed) Key environments. Oxford Pergamon Press, Galapagos, pp 145–156
Duffie CV, Glenn TC, Hagen C, Parker PG (2008) Microsatellite markers isolated from the flightless cormorant (*Phalacrocorax harrisi*). Mol Ecol Resour 8:625–627
Duffie CV, Glenn TC, Vargas FH, Parker PG (2009) Genetic structure within and between island populations of the flightless cormorant (*Phalacrocorax harrisi*). Mol Ecol 18:2103–2111
Edgar GJ, Banks S, Fariña M, Calvopiña M, Martínez C (2004) Regional biogeography of shallow reef fish and macro-invertebrate communities in the Galapagos archipelago. J Biogeogr 31:1107–1124
Eibl-Eibesfeldt I (1958) The Galapagos Islands: a laboratory of evolution. New Sci 4:250–253
Eibl-Eibesfeldt I (1984a) The large iguanas of the Galapagos Islands. In: Perry R (ed) Key environments: Galapagos. Pergamon Press, Oxford, pp 157–173
Eibl-Eibesfeldt I (1984b) The Galapagos seals. Part 1. Natural history of the Galapagos sea lion (*Zalophus californianus wollebaeki*, Sivertsen). In: Perry R (ed) Key environments: Galapagos. Pergamon Press, Oxford, pp 207–214
Eliasson U (1984) Native climax forest. In: Perry R (ed) Key environments: Galapagos. Pergamon Press, Oxford, pp 101–114
Ferodov AV, Dekens PS, McCarthy M, Revelo AC, de Menocal PB, Barreiro M, Pacanowski RC, Philander SG (2006) The Pliocene paradox (mechanisms for a permanent El Niño). Science 312:1485–1489
Fessl B, Kleindorfer S, Tebbich S (2006) An experimental study on the effects of an introduced parasite in Darwin's finches. Biol Conserv 127:55–61
Geist D (1996) On the emergence and submergence of the Galapagos Islands. Not Galapagos 56:5–9
Geist D (2009) Islands on the move: significance of hotspot volcanoes. In: de Roy T (ed) Galapagos: preserving Darwin's legacy. David Bateman Ltd, New Zealand, pp 28–35
Gentile G, Fabiani A, Marquez C, Snell HL, Snell HM, Tapia W, Sbordoni V (2009) An overlooked pink species of land iguana in the Galapagos. Proc Natl Acad Sci 106:507–511
Glynn PW (1986) Ecological effects of the 1982/83 El Niño Associate Disturbance to Eastern Pacific coral reefs. Progress Report 1986. CDRS Library, pp 1–13
Glynn PW (1994) State of coral reefs in the Galapagos Islands: natural vs. anthropogenic impacts. Mar Pollut Bull 29:131–140
Gottdenker N, Walsh T, Vargas H, Duncan M, Merkel J, Jimenez G, Miller RE, Dailey M, Parker PG (2005) Assessing the risks of introduced chickens and their pathogens to native birds in the Galapagos Archipelago. Biol Conserv 126:429–439
Grant PR (1984) The endemic land birds. In: Perry R (ed) Key environments: Galapagos. Pergamon Press, Oxford, pp 175–189
Grant PR (1999) Ecology and evolution of Darwin's finches, 2nd edn. Princeton University Press, Princeton, NJ
Grant BR, Grant PR (1989) Evolutionary dynamics of a natural population: the large cactus finch of the Galapagos. University of Chicago Press, Chicago
Grant PR, Grant BR (2006) Evolution of character displacement in Darwin's finches. Science 313:224–226
Grant PR, Grant BR (2008) How and why species multiply: the radiation of Darwin's finches. Princeton University Press, Princeton, NJ
Grove J (1985) Influence of the 1982–1983 El Niño event upon the icthyofauna of the Galapagos archipelago. In: Robinson G, Del Pino E (eds) El Niño en las Islas Galapagos: Evento 1982–1993. Charles Darwin Foundation, Galapagos, Ecuador, pp 245–258
Harris MP (1984) Seabirds. In: Perry R (ed) Key environments: Galapagos. Pergamon Press, Oxford, pp 191–206
Herbold R, Moyle PB (1986) Introduced species and vacant niches. Am Nat 128:751–760
Hoeck PEA, Bollmer JL, Parker PG, Keller LF (2010) Differentiation with drift: a spatio-temporal genetic analysis of Galapagos mockingbird populations (*Mimus* spp.). Phil Trans R Soc B 365:1127–1138

Houvenaghel GT (1984) Oceanographic setting of the Galapagos Islands. In: Perry R (ed) Key environments: Galapagos. Pergamon Press, Oxford, pp 225–231

Hull JM, Savage WK, Bollmer JL, Kimball RT, Parker PG, Whiteman NK, Ernest HB (2008) On the origin of the Galapagos hawk: an examination of phenotypic differentiation and mitochondrial paraphyly. Biol J Linn Soc 95:779–789

Huxley JS (1938) Species formation and geographic isolation. Proc Linn Soc Lond 150:253–264

Jiménez-Uzcátegui G, Carrión V, Zabala J, Buitrón P, Milstead B (2008) Status of introduced vertebrates in Galapagos. Galapagos report 2006–2007. Charles Darwin Foundation, Puerto Ayora, pp 136–141

Jiménez-Uzcátegui G, Carrión V, Zabala J, Buitrón P, Milstead B (2008) Status of introduced vertebrates in Galapagos. Galapagos report 2007–2008. Charles Darwin Foundation, Puerto Ayora, pp 97–102

Jordan MA, Snell HL (2008) Historical fragmentation of islands and genetic drift in populations of Galapagos lava lizards (Microlophus albemarlensis complex). Mol Ecol 17:1224–1237

Jordan MA, Hammond RL, Snell HL, Jordan WC (2002) Isolation and characterization of microsatellite loci from Galapagos lava lizards (Microlophus spp.). Mol Ecol Notes 2:349–351

Kennedy M, Valle CA, Spencer HG (2009) The phylogenetic position of the Galápagos Cormorant. Mol Phylogenet Evol 53:94–98

Kizirian D, Trager A, Donnelly MA, Wright JW (2004) Evolution of Galapagos Island lava lizards (Iguania: Tropiduridae: Microlophus). Mol Phylogenet Evol 32:761–769

Lack D (1947) Darwin's finches. Cambridge University Press, Cambridge

Lambeck K, Chappell J (2001) Sea level change through the last glacial cycle. Science 292:679–686

Lande R (1980) Genetic variation and phenotypic evolution during allopatric speciation. Am Nat 116:463–479

Lande R, Barrowclough GF (1987) Effective population size, genetic variation, and their use in population management. In: Soule ME (ed) Viable populations for conservation. Cambridge University Press, New York, pp 87–123

Larson EJ (2001) Evolution's workshop: God and science on the Galapagos Islands. Basic Books, New York

Laurie A (1985) Santa Fe news letter. Noticias de Galapagos 41:20–21

Lea DW, Pak DK, Belanger CL, Spero HJ, Hall MA, Shackleton NJ (2006) Paleoclimate history of Galapagos surface waters over the last 135,000 yr. Quaternary Sci Rev 25:1152–1167

Levin I, Outlaw DC, Vargas FH, Parker PG (2009) Plasmodium blood parasite found in endangered Galapagos penguins (Spheniscus mendiculus). Biol Conserv 142:3191–3195

Limberger D (1985) El Niño on Fernandina. In: Robinson G, Del Pino E (eds) El Niño en las Islas Galapagos: Evento 1982–1993. Charles Darwin Foundation, Galapagos, Ecuador, pp 245–258

Lopez TJ, Hauselman ED, Maxson LR, Wright JW (1992) Preliminary analysis of phylogenetic relationships among Galapagos Island lizards of the genus Tropidurus. Amphibia-Reptilia 13:327–339

Mace GM, Lande R (1991) Assessing extinction threats: toward a reevaluation of IUCN threatened species categories. Conserv Biol 5:148–157

Mayr E (1954) Change in genetic environment and evolution. In: Huxley J, Hardy AC, Ford EB (eds) Evolution as a process. Allen & Unwin, London, pp 157–180

Mayr E (1982) The growth of biological thought: diversity, evolution, and inheritance. Harvard University Press, Cambridge

McArthur RH, Wilson EO (1967) The theory of island biogeography. Princeton University Press, Princeton, NJ

McBirney AR, Williams H (1969) Geology and petrology of the Galapagos Islands. In: Bowman RI (ed) The Galapagos, Proceedings of the Symposia G.I.S.P. University of California Press, Berkeley, pp 65–70

Milinkovitch MC, Monteyne D, Gibbs JP, Fritts TH, Tapia W, Snell HL, Tiedemann R, Caccone A, Powell JR (2004) Genetic analysis of a successful repatriation program: giant Galapagos tortoises. Proc R Soc Lond B 271:341–345

Miller GD, Hofkin BV, Snell H, Hahn A, Miller RD (2001) Avian malaria and Marek's disease: potential threats to Galapagos penguin *Spheniscus mendiculus*. Mar Ornithol 29:43–46

Palmer CE, Pyle RL (1966) The climatological setting of the Galapagos. In: Bowman RI (ed) The Galapagos, Proceedings of the Symposia G.I.S.P. University of California Press, Berkeley, pp 65–70

Parent CE, Crespi BJ (2006) Sequential colonization and diversification of Galapagos endemic land snail genus Bulimulus (Gastropoda, Stylommatophora). Evolution 60:2311–2328

Parent CE, Caccone A, Petren K (2008) Colonization and diversification of Galapagos terrestrial fauna: a phylogenetic and biogeographical synthesis. Phil Trans R Soc B 363:3347–3361

Parker P (2009a) A most unusual hawk: one mother and several fathers. In: de Roy T (ed) Galapagos: preserving Darwin's legacy. David Bateman Ltd., New Zealand, pp 130–137

Parker PG (2009b) Parasites and pathogens: threats to native birds. In: de Roy T (ed) Galapagos: preserving Darwin's legacy. New Zealand, David Bateman Ltd, pp 177–183

Parker PG, Whiteman NK, Miller RE (2006) Perspectives in ornithology: conservation medicine in the Galapagos Islands: partnerships among behavioral, population and veterinary scientists. Auk 123:625–638

Peck SB (1996) Origin and development of an insect fauna on a remote archipelago: the Galapagos Islands, Ecuador. In: Keast A, Miller SE (eds) The origin and evolution of Pacific Island biotas, New Guinea to Eastern Polynesia: patterns and processes. Academic Publishing, Amsterdam, pp 91–122

Petren K, Grant BR, Grant PR (1999) A phylogeny of Darwin's finches based on microsatellite DNA length variation. Phil Trans R Soc B 266:321–329

Petren K, Grant PR, Grant BR, Keller LF (2005) Comparative landscape genetics and the adaptive radiation of Darwin's finches: the role of peripheral isolation. Mol Ecol 14:2943–2957

Porter DM (1976) Geography and dispersal of Galapagos Islands vascular plants. Nature 264:745–746

Porter D (1984) Endemism and evolution in terrestrial plants. In: Perry R (ed) Key environments: Galapagos. Pergamon Press, Oxford, pp 85–99

Quiroga D (2009) Galapagos, laboratorio natural de la evolucion: una aproximacion historica. In: Tapia W, Ospina P, Quiroga D, Gonzales JA, Montes C (eds) Ciencia para la sostenibilidad en Galapagos. Parque Nacional Galapagos, Ecuador

Rassmann K (1997) Evolutionary age of the Galapagos iguanas predates the age of the present Galapagos Islands. Mol Phylogenet Evol 7:158–172

Riedinger M, Steinitz Kannan M, Last W, Brenner M (2002) A 6100 14-C record of El Niño activity from the Galapagos Islands. J Paleolimnol 27:1–7

Robinson G, del Pino EM (1985) El Niño in the Galapagos Islands. Charles Darwin Foundation, Quito

Roque-Albelo L (2008) Evaluating land invertebrate species: prioritizing endangered species. Galapagos report 2006–2007. Charles Darwin Foundation, Puerto Ayora, 111–117

Santiago-Alarcon D, Whiteman NK, Ricklefs RE, Parker PG, Valkiunas G (2008) Patterns of parasite abundance and distribution in island populations of Galapagos endemic birds. J Parasitol 94:584–590

Sato A, Tichy H, O'Huigin C, Grant PR, Grant BR, Klein J (2001a) On the origin of Darwin's finches. Mol Biol Evol 18:299–311

Sato A, Mayer WE, Tichy H, Grant PR, Grant BR, Klein J (2001b) Evolution of Mhc class II B genes in Darwin's finches and their closest relatives: birth of a new gene. Immunogenetics 53:792–801

Schluter D (1986) Character displacement between distantly related taxa? Finches and the bees in the Galapagos. Am Nat 127:95–102

Schluter D, Grant PR (1984) Determinants of morphological patterns in communities of Darwin's finches. Am Nat 123:175–196

Schluter D, Price TD, Grant PR (1985) Ecological character displacement in Darwin's finches. Science 227:1056–1059

Schmitz P, Cibois A, Landry B (2007) Molecular phylogeny and dating of an insular endemic moth radiation inferred from mitochondrial and nuclear genes: the genus Galagete (Lepidoptera: Autostichidae) of the Galapagos Islands. Mol Phylogenet Evol 45:180–192

Sequeira AS, Lanteri AA, Scataglini MA, Confalonieri VA, Farrell BD (2000) Are flightless *Galapaganus* weevils older than the Galapagos Islands they inhabit? Heredity 85:20–29

Shen GT, Cole JE, Lea DW, Linn LJ, McConnaughey TA, Fairbanks RG (1992) Surface ocean variability at Galapagos from 1936–1982: calibration of geochemical tracers in corals. Paleoceanography 7:563–588

Simkin T (1984) Geology of Galapagos Islands. In: Perry R (ed) Key environments: Galapagos. Pergamon Press, New York, pp 15–41

Simkin T, Howard KA (1970) Caldera collapse in the Galapagos Islands, 1968. Science 169:429–437

Snell HL, Tye A, Causton CE, Bensted-Smith R (2002) Current status and threats to the terrestrial biodiversity of Galapagos. A biodiversity vision for the Galapagos Islands. Charles Darwin Foundation and World Wildlife Fund, Puerto Ayora, Galapagos, pp 30–47

Sulloway FJ (1982) Darwin and his finches: the evolution of a legend. J Hist Biol 15:1–53

Swanson FJ, Baitis HW, Lexa J, Dymond J (1974) Geology of Santiago, Rabida and Pinzon Islands, Galapagos. Geol Soc Am Bull 85:1803–1810

Tapia W, Patry M, Snell H, Carrión V (2000) Estado actual de los vertebrados introducidos a las islas Galapagos. Fundación Natura: Informe Galapagos 1999–2000, Quito, Ecuador

Trillmich F (1984) Part 2. Natural history of the Galapagos fur seal (Arctocephalus galapagoensis, Heller). In: Perry R (ed) Key environments: Galapagos. Pergamon Press, Oxford, pp 85–99

Trillmich F (1985) Effects of the 1982/83 El Niño on Galapagos fur seals and sea lions. Not Galapagos 42:22–23

Trillmich F, Limberger D (1985) Drastic effect of El Niño on Galapagos, Ecuador pinnipeds. Oecologia 67:19–22

Tye A (2007) The status of the endemic flora of Galapagos: the number of threatened species is increasing. Galapagos report 2006–2007. Charles Darwin Foundation, Puerto Ayora, pp 97–103

Tye A (2008) The status of the endemic flora of Galapagos: the number of threatened species is increasing. Galapagos report 2007–2008. Charles Darwin Foundation, Puerto Ayora, pp 97–102

Tye A, Snell HL, Peck SB, Andersen H (2002) Outstanding terrestrial features of the Galapagos Archipelago. A biodiversity vision for the Galapagos Islands. Charles Darwin Foundation and World Wildlife Fund, Puerto Ayora, Galapagos, pp 12–23

Tye A, Atkinson R, Carrión V (2008) Increase in the number of introduced plant species in Galapagos. Galapagos report 2006–2007. Charles Darwin Foundation, Puerto Ayora, pp 133–135

Urban FE, Cole JE, Overpeck JT (2000) Influence of mean climate change on climate variability from a 155-year tropical Pacific coral record. Nature 407(6807):989–993

Valle CA (1985) Alteración de las poblaciones del cormorán no volador, el pingüino y otras aves marinas en Galapagos por efecto de El Niño 1982–83 y su subsecuente recuperación. In: Robinson G, Del Pino E (eds) El Niño en las Islas Galapagos: Evento 1982–1993. Charles Darwin Foundation, Galapagos, Ecuador, pp 245–258

Valle CA (1994) Pepino war, 1992—is conservation just a matter for the elite? Not Galapagos 53:2

Valle CA, Coulter MC (1987) Present status of the flightless cormorant, Galapagos penguin, and greater flamingo populations in the Galapagos Islands, Ecuador after the 1982–83 el Niño. Condor 89:276–281

Valle CA, Cruz F, Cruz JB, Merlen G, Coulter MC (1987) The impact of the 1982–1983 El Niño Southern oscillation on seabirds in the Galapagos Islands, Ecuador. J Geophys Res 92:14437–14443

Van Riper C, Van Riper SG, Goff ML, Laird M (1988) The epizootiology and ecological significance of malaria in Hawaiian land birds. Ecol Monogr 58:111–127

Vargas H, Snell HM (1997) The arrival of Marek's disease to Galapagos. Not Galapagos 58:4–5

Vargas FH, Harrison S, Rea S, Macdonald DW (2006) Biological effects of El Niño on the Galapagos penguin. Biol Conserv 127:107–114

Whiteman NK, Goodman SJ, Sinclair BJ, Walsh T, Cunningham AA, Kramer LD, Parker PG (2005) Establishment of the avian disease vector *Culex quinquefasciatus* Say, 1823 (Diptera: Culicidae) on the Galapagos Islands, Ecuador. Ibis 147:844–847

Whiteman NK, Kimball RT, Parker PG (2007) Co-phylogeography and comparative population genetics of the Galapagos Hawk and three co-occurring ectoparasite species: natural history shapes population histories within a parasite community. Mol Ecol 16:4759–4773

Wikelski M, Foufopoulos J, Vargas H, Snell H (2004) Galapagos birds and diseases: invasive pathogens as threats for island species. Ecol Soc 9:5

Williams H (1966) Geology of the Galapagos Islands. In: Bowman RI (ed) The Galapagos, Proceedings of the Symposia G.I.S.P. University of California Press, Berkeley, pp 65–70

Williams H, McBirney AR (1979) Volcanology. Freeman, Cooper, San Francisco

Withman JD, Smith F (2003) Rapid community change at a site in the Galapagos marine reserve. Biodivers Conserv 12:25–45

Wright S (1931) Evolution in Mendelian populations. Genetics 16:97–159

Wright S (1932) The roles of mutation, inbreeding, crossbreeding and selection in evolution. Proc 6th Int Cong Genet 1:356–366

Wyles JS, Sarich VM (1983) Are the Galapagos iguanas older than the Galapagos? Molecular evolution and colonization models for the archipelago. In: Bowman RI, Berson M, Levinton AE (eds) Patterns of evolution in Galapagos organisms. Amer. Assoc. Advanc. Sc., Pacific Div, San Francisco, pp 177–185

Wyrtki K (1975) 'El Niño': the dynamic response of the Equatorial Pacific Ocean to atmospheric forcing. J Phys Oceanogr 5:572–584

Zachos J, Pagani M, Sloan L, Thomas E, Billups K (2001) Trends, rhythms, and aberrations in global climate 65 Ma to present. Science 292:686–693

# Chapter 2
# Changing Views of the Galapagos

Diego Quiroga

## Introduction

Social constructs emerge in specific economic and sociopolitical contexts and are associated with particular groups with concrete interests and histories and under particular regimes (Proctor 1998). As Escobar (1994) has indicated, many of the constructs and categories used to understand the world have been produced in developed countries and are being used and exported to the rest of the world. Escobar talks about the problematization of specific issues and the way in which international bureaucracies use discourses to create professionals and experts that can solve issues such as poverty, malnutrition, and environmental degradation, which have been framed in particular ways. As Escobar and other authors have reminded us, there is an economic system that supports these constructs and perceptions.

In the process of the expansion and conquest of new areas, discourses may encounter alternative and incommensurable framings and definitions. As defined by Elizabeth Povinelli, incommensurability refers to a "state in which undistorted translation cannot be produced between two or more denotational texts" (Povinelli 2001, p. 329). These encounters produce different results that range from the coexistence of two frameworks in relative isolation—in the case of heterodoxic societies—to the absorption of one framework by the other, as in orthodoxic societies (Bourdieu 1984). Espeland and Mitchell (1998) have pointed out the ways in which bureaucracies create orthodoxy as they depend on the standardization between disparate things that reduces the relevance of context. This process which is termed commensuration consists of reducing the difference and the generation of consensus.

D. Quiroga (✉)
College of Biological and Environmental Sciences, Universidad San Francisco de Quito,
Quito, Ecuador
e-mail: dquiroga@usfq.edu.ec

S.J. Walsh and C.F. Mena (eds.), *Science and Conservation in the Galapagos Islands:*     23
*Frameworks & Perspectives*, Social and Ecological Interactions in the Galapagos Islands 1,
DOI 10.1007/978-1-4614-5794-7_2, © Springer Science+Business Media, LLC 2013

Using this theoretical framework, I will illustrate the way in which specific constructs of nature have been generated in the Galapagos at different times and by different groups as they have come to the islands and how these constructs interact to generate new and hybrid understandings. In the Galapagos, several authors (Ospina 2006; Grenier 2007; Quiroga 2009a, b; Hennessy and McCleary 2011) have recently explored the interactions between different groups of people, their specific and concrete activities, and their constructs and models. From this analysis, it is clear that the global conception of the Galapagos is one that views the islands as a perfect place where nature can be studied and key evolutionary processes understood. To a large extent, the basis for this construct is the idea popularized by Charles Darwin and other early scientists that the Galapagos constitutes an ideal natural laboratory.

The history of encounters in the Pacific includes many examples where incommensurable visions have encountered each other. European expansion in the Pacific is filled with these encounters between incommensurable visions (Sahlins 1995; Obeyesekere 1997). One of the best examples is that of the fatal encounter between Captain James Cook and the native people of Hawaii. As described by Marshal Sahlins, there are a series of incongruities between the two theoretical approaches. Similarly, Margaret Jolly and Serge Tcherkezoff (Howes 2011) have described the incommensurability between the concepts of the native people living in the Pacific Islands and the Europeans during the European exploration and conquest of these islands, including misinterpretations of sexual encounters and power relations (Tcherkezoff 2009). In these cases, we can talk about incommensurable world views, as the understanding of nature, spirituality, and the Other was based on ideas and concepts that were fundamentally different (Povinelli 2001). In the case of the Galapagos, the encounter was not between native islanders and the European explorers and scientists but between a later group of European explorers and scientists like Charles Darwin and Robert Fitz Roy, who were to a large extent following a tradition started by previous explorers of the Pacific, like Cook, d'Entrecasteaux, Bougainville, and the Ecuadorian colonists. Although, in many cases, the two groups that encountered each other in the Galapagos were much closer in their perceptions and basic conceptual understandings than those of other part of the Pacific, the differences between the two paradigms were important enough to justify the qualification of being incommensurable.


## The Scientific View


The importance of the Galapagos Islands for the development and testing of different, and often contrasting, ideas about the evolution of species emerged early in the nineteenth century with Charles Darwin's visit. The debates and clashes surrounding Darwin's ideas became the foundations for the construction of the islands as a natural laboratory. Later, this construct shaped other visions of the Galapagos, such as those produced by conservationists, the tourism sector, and, increasingly, the conceptual framings of the local residents.

There are several biogeographical reasons why the Galapagos has been considered a natural laboratory for the study of evolution: the distance between the islands and the mainland, which provides some degree of isolation that results in the evolution of the different species, and the age of the islands, for if the islands were much younger, then species would not have had time to diverge, but if they were much older, the species would be so different that they would have been more difficult to recognize as evolving from same species. The fact that there was no early colonization of the islands by pre-Hispanic people explains to some extent why more than 90% of the endemic animals are still there (Valle and Parker 2012). The diversity provided by the currents and the different altitudinal ecological zones makes the Galapagos a particularly interesting place to study evolution. Furthermore, the Galapagos being tropical islands has an unusually rich and dynamic marine environment, the result of a series of oceanic currents that give scientists an opportunity to watch populations adapt to changes in a relatively short time.

Darwin was the first visitor to the islands to develop a concrete and coherent explanation relating geological, geographic, and biological aspects and, thus, initiating the modern science of biogeography and evolutionary biology. Despite the scientific importance of Darwin's visit to the development of his theory, the visit of the *HMS Beagle* to the Galapagos also has elements of a modern secular myth. As has been shown by Sulloway (1982, 1984) and other authors, Darwin's supposedly instant conversion to evolutionism away from creationism in the Galapagos never occurred. This secular myth points to the Galapagos as the place where Darwin had his revelation and his major insights. Thus, in the popular history of evolutionary science, the Galapagos has become a kind of Mecca of evolution, a place where one can observe, as Darwin did, the processes and mechanisms at work (Hennessy and McCleary 2011). Sulloway (1982) has indicated that far from being a specific eureka moment, it was not a single eureka-type discovery based on Darwin's observation of the finches, but rather, it was a long process of analysis and reflection, and it was not the finches but rather the mocking birds that made Darwin consider the possibility of the existence of the transmutation of the species. The biological bases for the differences between the two types of birds lie in the fact that the mocking birds, due to their territoriality and reproductive patterns, are much less likely to move from island to island than the finches, and are represented by four different species, three of which are characteristic to a particular island. The distribution of these birds made it possible for Darwin, who collected three of the four species, to start thinking about the transmutation of species. The distribution of the mocking bird species and the small differences between species living on different islands was one of the facts that Darwin eventually noticed that forced him to raise key questions about the origin of species. The fossils that Darwin saw in South America and the mocking birds he saw in the Galapagos and in Chile, as has been discussed by Durham (2012), created important anomalies that the previous paradigms could not explain. Durham points out that there were two types of anomalies with which Darwin was struggling. One was the affinity anomaly which refers to the similarity between biota of oceanic islands and neighboring continental islands, and the second is the replacement anomaly concerning the way in which similar species appear to succeed one another in time or take each other's place in nature (Durham 2012). We do know that Darwin begins to question the creationist view during the last part of his almost 5-year trip around the world in the *Beagle* (Sulloway 1984). Influenced by thinkers like Thomas Malthus and Charles Lyell, he develops a gradualist view that sees continuous change as the norm. Once he is back in England, Darwin's observations about the differences between species of birds, such as the finches, and reptiles, such as the tortoises, benefit from the help of leading ornithologists like John Gould and become important elements in the development of the idea of species evolution by natural selection (Sulloway 1982; Durham 2012). These anomalies, which were few but fundamental and that indicated for Darwin the possibility of the transmutation of the species, generated a process that resulted in one of the most important paradigm shifts of modern times. It is this revolution that put Darwin and the Galapagos at the epicenter of the debates and studies that followed the publication of *On the Origin of Species* in 1859.

For Darwin, one of the main lessons for the study of evolution that the Galapagos and other oceanic islands could provide had to do with the distribution of the species and their dispersal. After Darwin, many scientists realized that the Galapagos acted as a living museum where evolutionary patterns could be understood (Quiroga 2009a, b; Hennessy and McCleary 2011). As is well known, however, Darwin's ideas initiated a long debate in the nineteenth century and at the beginning of the twentieth century as many biologists rejected the conclusions that Darwin had reached. One of the most charismatic of these biologists was Louis Agassiz, a Swiss-born Harvard professor, who was a creationist and a catastrophist. He believed that mutations can only create monstrosities and he indicated that "All such facts seem to show that the so-called varieties or breeds, far from indicating the beginning of new types, or the initiating of incipient species, only point out the range of flexibility in types which in their essence are invariable" (Agassiz 1896). For him, the distribution of the species in places like the Galapagos and the Amazon River proved that Darwin was wrong for, Agassi reasoned, how else could one explain that in similar environments and climates, species could be so different (Agassiz 1896; Winsor 1979; Dexter 1979; Morris 1988). For Agassiz, the Galapagos served as one of the scenarios that he hoped could discredit the ideas of Darwin (Larson 2001). A few years before his death, Agassiz sailed in the *Hassler* to the Galapagos as part of his campaign to discredit Darwinism.

The triumph of Darwinism in the biological sciences has resulted in Darwin becoming an important icon for science and for popular culture. The Darwin secular myth (his travels and his life) has many elements that equate him to a religious figure. As is the case with many mythical religious and secular figures, Darwin's trip on the *Beagle* is a hero's journey, a time of hardship but also of revelation. The modern, secular view of the evolution of life on Earth that now prevails in a large part of the Western population is, in part, the result of Darwin's observations in the Galapagos, as he later admits in his journals. It is based on the idea that Darwinian processes unregulated, random, and undirected generate an order, albeit an imperfect one, by the very nature of their emergent properties. Imperfect complex forms, such as the ones that exist in nature, result from a simple set of key rules, such as the generation of diversity and the natural selection of the fittest forms. The 5 weeks that Darwin spent and the four islands he visited on the Galapagos were very important in initiating this profound paradigm shift.

For Darwin and for many other evolutionists, the importance of the Galapagos depends to a large extent on its isolation from the mainland. The isolation of the islands from the mainland was not always assumed as a fact, and scientists during the nineteenth century, in particular Baur, have maintained that the islands were at some point connected to the mainland (Baur 1891). Once the idea of the isolation of the islands and the Darwinian paradigm of evolution were widely accepted at the beginning of the twentieth century (Larson 2001), scientists like David Lack and Peter and Rosemary Grant based their studies on the use of the isolation of the islands to understand the evolution of the species (Grant 2008). It is within the theoretical framework of Darwinian evolution and the fact that the islands are of volcanic origin that the Galapagos starts to become famous as a natural laboratory for the study of evolution.

The connection with the Galapagos and the study of evolution does not of course end in the early twentieth century; the relevance of the Galapagos today derives from hundreds of meticulous studies such as the Grants' research on finches, Duncan Porter's work on plant evolution and distribution, Guy Coppois' incredible example of adaptive radiation with the bulimulid land snails, and Gisella Caccone's research on the distribution and evolution of tortoises. As new techniques and methods such as genetic, studies, GIS, and mathematical modeling in ever more powerful computers became available, the Galapagos became a referent on this side of the Atlantic (Quiroga 2009b). The Galapagos is one of those remarkable places that provide an ideal scenario where many Darwinian evolutionists can test their ideas. Evolutionists concerned with fitness peaks and valleys can use this scenario to better understand the distribution of genetic characteristics on the different islands. Genetics, statistics, and computer power are now used to test models and ideas in this natural laboratory (Valle and Parker 2012).

Conservationist concerns, as we understand them today, have not always been associated with scientific sensitivities. Concerns about the health of the Galapagos were already expressed by scientists in the nineteenth and early twentieth centuries, but in the early days, these concerns translated mostly into an effort to collect specimens from the Galapagos in order to save them from being lost to science. In 1907, eight young scientists chosen and sent by the California Academy of Sciences (CAS) went to the Galapagos on the 89-foot schooner *Academy*. The expedition led by Rollo Beck spent a year collecting on all the major and minor islands of the Galapagos (James 2010). The fear that the animals would be gone within a few years—a concern that had been expressed before by previous scientist–collectors such as Albert Gunter and Walter Rothschild—motivated the CAS expedition to collect 75,000 biological specimens, more than any expedition to the islands before and since (James 2010). It brought over 260 preserved specimens of giant tortoises as well as numerous specimens of reptiles, birds, mammals, insects, plants, land snails, and fossils (James 2010). These efforts to collect reflect the idea, common at the time, that collecting was the only way of safely preserving and studying the specimens (James 2010). Huge collections, such as those of the CAS, are in part responsible for the fame of the islands as Darwin's living outdoor laboratory of evolution (James 2010).

In the 1930s and 1940s, a new view of conservation and protection in situ of the fauna and flora of the islands was developed by researchers like Austrian ethologist Irenäus Eibl-Eibesfeldt. Eibl-Eibesfeldt's idea was to conserve the animals living on the islands for future generations. In 1933, German naturalist Victor Von Hagen started to promote his project to commemorate 100 years of the *Beagle* and proposed the creation of a scientific station. His idea was not immediately accepted as there were more grave concerns occupying the politicians and at the time scientists were just starting to accept Darwin's ideas as a universal paradigm (Ospina 2004). Von Hagen was the main proponent of the idea that several of the islands be declared a Fauna Reserve in May 1936. But the Second World War made it impractical to really establish the reserve, and only one guard was assigned to it. As Darwin and Darwinism became the dominant paradigm of the scientific community and as the genetic synthesis fused Mendelian genetics and the Darwinian theory of natural selection, some of the leaders of this new perspective such as Ernst Mayr and Julian Steward pressed for the protection of the islands (Ospina 2004; Hennessy and McCleary 2011). Julian Huxley, a very influential and powerful person in the scientific community, was one of the early proponents of schemes to conserve the Galapagos. As the grandson of Thomas Huxley—a man known as Darwin's bulldog because of his aggressive defense of Darwinism—Julian had a personal and philosophical interest in defending the Galapagos, and he turned the protection of the islands into a personal crusade (Larson 2001). He believed in the evolutionary progress of the human mind from lower to higher forms. He was a prominent supporter of eugenics and the use of science to allow the preferential breeding of the best of humankind (Cairns 2011). In 1946, after the Second World War, he was elected as the first general director of UNESCO, and immediately afterward he persuaded the organization to include conservation to its agenda. He convinced UNESCO that the Galapagos should be a key conservation site in part because of his links to the history of Charles Darwin, and he was a key figure in the declaration of the Galapagos as a national park and the creation of the Charles Darwin Foundation (CDF) (Cairns 2011). In 1954, while president of the Royal Society, Huxley supported the visit to the Galapagos of a mission led by Eibl-Eibesfeldt and American zoologist Robert Bowman due to his concerns about scientists' complaints regarding the possible negative effects that the 2,000 residents might have on the Galapagos (Larson 2001). It was this visit that resulted in the creation of the CDF (Cairns 2011). This scientific vision of the Galapagos started becoming popular in the 1960s with a series of television, magazine, and film productions (Hennessy and McCleary 2011). Because some early proponents of the idea of the Galapagos becoming a conservation sanctuary had strong ecocentric views, some authors have speculated that an ecofascist vision was guiding these early views of the islands, views proposed mostly by foreigners and outsiders (Orduna 2008) that were very critical of the residents of the Galapagos who they perceived mostly as a threat to the islands' biodiversity. As the Galapagos became better known for being a natural laboratory and a place where scientists could study and understand evolution, local people were perceived as a disruptive force that needed to be dealt with. Thus, a dominant view was established that conceived of the Galapagos as an ideal natural laboratory, due to their basic geological, geographic, and biological characteristics, and that viewed the local people as a menace to conservation and to the maintenance of the Galapagos and its uniqueness (Quiroga 2009a, b; Hennessy and McCleary 2011).

Starting in the 1970s, a new economic and discursive activity started to flourish in the islands. The influx of tourists to the Galapagos in general, and the growth of tourists staying in accommodations in the towns specifically, played a key role in the creation of the new hybrid discourse and increased commensurability between the value systems. The Galapagos Islands provide a series of physical, biological, and cultural conditions that make them attractive to international visitors. Some of these include the tameness of the fauna, which are easily approached by visitors, the iconic aspects of the islands, the existence of emblematic species, and the increasingly better infrastructure and amenities, such as 24-h electricity, food refrigeration, air conditioning, fast boats, restaurants, and better communications (Grenier 2007; Quiroga et al. 2010). Furthermore, as in many other destinations (Becken 2010), the sense of safety and a favorable climate play an important role in attracting the large number of tourists to visit each year.

From the middle of the twentieth century, conservationists saw tourism as way to protect the biodiversity of the islands (Grenier 2007; Ospina 2001). Businessmen from mainland Ecuador and from developed nations and locals from the Galapagos have used the idea of the islands as a pristine natural laboratory to create a multimillion dollar industry. As has been mentioned by several authors (Ospina 2001; Grenier 2007; Quiroga 2009a, b; Hennessy and McCleary 2011), tourism constitutes the appropriation and commercialization of the global discourse about the Galapagos.

Tourism, especially large tourism operations, shares with the conservation sector much of the discourse of saving nature from extractive activities. Many of the owners and operators of large vessels frequently mention the necessity of protecting the Galapagos from the destructive hands of the local population. From their beginning, large tourism operations were planned and programmed as activities that should serve the conservation effort. In a report from 1957, a UNESCO reconnaissance mission suggested that the Galapagos could become an important asset for the Ecuadorian economy by attracting tourism. The 1966 Snow and Grimwood Report recommended ways in which tourism could be managed by large companies (Cairns 2011). The use of floating hotels was to play a key role in the process (Grenier 2007; Cairns 2011). To a large extent, this so-called floating hotel model, which many now agree has backfired, was based on a perception that the local population was the main problem for the conservation of the islands. This new view of the Galapagos originated from the recognition that the biodiversity that exists in the islands is the main resource to be utilized in a non-extractive and sustainable manner. Thus, the imposition of this agenda and the creation of floating hotels resulted in the consolidation of an alliance between cruise boat tourism, science, and conservation (Grenier 2007; Ospina 2001).

Tourism infrastructure is concentrated in the hands of a few people who often have important connections with the conservation sector and share the same visions and concerns. Taylor et al. (2006) have indicated that in 2005, foreigners and mainland residents owned most of the top level luxury boats (almost 82% of them), while Galapagos residents owned only 18%. On the other hand, Galapagos residents owned most of the economy class boats (73%). With some notable exceptions, the companies that own and operate the more expensive boats based mostly in Quito and Guayaquil (Taylor et al. 2006; Epler 2007) have traditionally shown little interest in the development of the local towns, as their operations have largely ignored the towns as part of the destination. The discourse produced by many of these operators and agencies emphasizes the Galapagos as a pristine land where people are absent and pristine nature can be observed (Grenier 2007). Pretending that the local people are invisible, as the cruise boat tourism chooses to do, achieves, at least in the plane of representation, what some scientists and conservationists wanted to achieve in practice. An Internet search for pages advertising tours of the Galapagos reveals an emphasis on tame and friendly animals; in most of the pages, there is no mention of the local inhabitants (Grenier 2007).

Although the dominant construct of the Galapagos produced by cruise boat tourism shares many of the basic concepts with the scientific constructs of the islands, it also differs from the scientific perspective in important ways. It is a simplified and domesticated view of Darwinism, as some of the most troubling Darwinian ideas have been packaged for popular consumption. As Ospina has noted, the Darwinian paradigm, which views a constant struggle between organisms for survival and considers diversity and natural selection as the main drivers of evolution, is often transformed by many involved in the tourism sector, such as tour operators, travel agencies, and guides, into a more harmonious view of nature in which tame creatures live in a peaceful manner and can be observed by humans (Ospina 2004, 2006).

## The Galapagos as a Frontier and Extractive Economies

The extraction of resources from the Galapagos started early, with tortoises being taken away by pirates and privateers and later by whalers and fur hunters. This idea of the Galapagos as a source of goods to be extracted continued during the time of colonization by Ecuadorians (González et al. 2008). Just a few years after Ecuador was created as a nation in 1830, it declared the islands part of its territory and made an effort to annex them. Before and after Ecuador annexed the Galapagos, several other countries were interested in the islands (among them the UK, the USA, and Chile), wishing either to extract their resources (products such as tortoises, whales, sea lions, guano, orchilla, fish) or to use them as a strategic geopolitical outpost. This idea appealed to and moved not only the young country of Ecuador but also countries such as the USA (Latorre 2001).

The first colonists arrived in the archipelago in the early 1830s, as Ecuador tried to establish control of the land. The first group led by Jose Villacis, a veteran of the wars of Andean independence from the Spaniards, established a colony in Floreana or Charles islands. In 1860, a second colony was established in San Cristobal, and, eventually, a sugar plantation and sugar mill, a coffee plantation, and cattle farm were constructed. As in the case of Floreana, this colony led by J. M. Cobos was based to a large extent on outlaws and political prisoners, some of whom eventually killed Cobos, who they accused of being their brutal oppressor. A similar pattern occurred in Isabela where cattle farms and plantations were also created. Once these colonies were dissolved, the people who stayed started to control and manage their own farms, or *fincas*. During the 1960s and 1970s, the Ecuadorian public's view of the Galapagos as a frontier—a remote and harsh place, where the land could be tamed through hard labor and the creation of agriculture and cattle farms—was further enhanced. As was the case of the Oriente (Ecuadorian Amazon forest), the Galapagos became a region of agricultural expansion and colonization. The Galapagos was conceived as the land of transformation from wild nature to culture—a land that humans through their labor could domesticate and control. Contrary to the Darwinian evolutionist classification of organisms as either endemic, native, or introduced, the early colonists saw animals as either useful, useless, or pests. Preserving the isolation—a requirement for the maintenance of the natural laboratory that scientists were dreaming of—was exactly what the locals and their economic logic were trying to avoid.

For many Ecuadorian colonists, nature must be conquered and the land *cleaned* (cutting the forest is often referred to as *limpiar el monte*). Areas like the Galapagos and the Amazon were subject to laws passed by developmentalist governments that promoted colonization. The Ecuadorian government needed to expand its frontiers in part as a response to pressure from poor people in the highlands who needed more land. Laws were passed during the twentieth century punishing those who kept the land idle, and conservation was neither a concern nor a priority. Much of this vision still permeates the views and desires of many Ecuadorians living in the rural areas of the Galapagos. Thus, in the case of Isabela, some residents still consider animals such as tortoises as sources of food and the Galapagos hawk as a pest that kills their chickens and needs to be eliminated.

According to this pioneer mentality, the transformation of wild nature into domesticated and productive nature is an act of possession and ownership (Ospina 2001). Many pioneers feel that through their labor and hardwork, they transformed the islands from a harsh and difficult place to one where people could live comfortably. They think of themselves as the ones who made the islands hospitable and that they have undisputable rights over the land and the seas that newcomers do not have. They remind younger residents, especially newcomers, that they created the basic infrastructure, such as the airport in San Cristobal and the roads. They even claim that the legal system that supports much of the environmental policies they dislike was of their making, as some fishermen say occurred with the creation of the Galapagos Marine Reserve (GMR). These early colonists are often called by the term *carapachudos* (from the Spanish word for carapace referring to the Galapagos tortoises) which, as Ospina points out, refers to their rough character and the fact that they can live without food or water and support the hardest conditions (Ospina 2001, p. 30). Unfortunately, they complain that their hard labor is now mostly benefiting others, especially outsiders who are now establishing their businesses on the islands.

The pioneers romanticize the past. It is described as a time when there were no diseases or problems with agricultural pests and when sea animals were plentiful and nature provided all the resources people needed. In the ocean, there were plenty of lobsters and fish; it was enough to go to the shore and collect all the sea animals they wanted or to go to the highlands and hunt the wild pigs and goats. "We used to go down to the shore," a 67-year-old fisherman told me, "and we would take as many lobsters as we wanted. Since we did not have a refrigerator, we used to take only those we needed, there was no need for any type of controls, and we never overfished." For them, that was the real Galapagos and that was real conservation. Most of the older people interviewed maintain that there were no environmental problems; those are to be blamed on the conservationists, the industrial fishing boats, and the large tourism companies. There was a high degree of isolation as, until the late 1950s, only one or two boats would come per year. The coming of a boat was an important event not only because it brought goods and letters from distant friends and relatives but also because people in San Cristobal who had not seen each other for several months used it as an opportunity to meet. It was a time of celebration.

The original agricultural sector became less predominant in the second half of the twentieth century as new sectors became leaders. The most important of these sectors was fishing, which started to grow in the 1950s. Colonists, who were originally dedicated to agriculture and lived in the highlands, began to descend to the beach areas and to participate in different fishing activities. Many were fishing for bacalao (*Myctoperca olfax*), which they salted and dried and sent to the mainland to be prepared as a soup to be eaten during Easter celebrations. Other fishermen captured fish, turtles, sharks, and lobsters that they sold to large industrial boats from different countries (especially Japan), which were anchored in San Cristobal's port. Green and red spiny lobsters fisheries (*Panulirus penicillatus* and *P. gracilis*) which started in the 1960s became major exports in the 1980s (Hearn 2008); most were sold to Guayaquil from where it was exported to the mainland. Lobster fishermen introduced the hookah system which consists of a compressor that provides air to a diver (Castrejón 2006). As Southeast Asian economies improved, there was increased demand for sea cucumbers (*Isostichopus fuscus*), and an emergent fishing industry was created in a short period of time during the early 1990s. The Galapagos National Park (GNP) tried to control the fishing industry in the middle of that decade, but that resulted in tensions and conflicts (Hearn 2008) (Castrejón 2006). Between 1995 and 2005, several strikes and conflicts paralyzed the GNP, creating instability and mismanagement (Hearn 2008; Quiroga 2009a, b). During this time, the extractive versus conservationist discourses and visions clashed constantly and became more polarized. The polarization made it clear that the conflict was not only between two economic conditions but also between two divergent and incommensurable cosmologies and valuations of nature.

Due to the increasing amounts of money that the Galapagos fisheries brought to local communities, the number of registered and active fishermen in the Galapagos increased from 752 in 1999 to 1,229 in 2000. More recently, however, due to the collapse of the main fisheries, the number of active fishermen decreased to 436 by

2007 (Castrejón 2006; Quiroga 2011), representing roughly 2% of the total population. The reduction of resources in the marine reserve, particularly sea cucumbers, demersal fish such as groupers, and lobster (although the latter has recovered in recent years), explains this decrease. Many fishermen have transitioned to more profitable sectors, such as tourism. Currently, the GMR consists of approximately 450 active and registered fishermen (Quiroga 2009a, b). By 2006, fishing made up less than 4% of local income (Watkins and Cruz 2007).

For many fishermen, the creation of the GMR in 1998 meant alienation and restrictions (Quiroga 2009a, b). Some fishermen feel that although they were the main promoters of the GMR, the reserve's regulations have in many ways benefited other people, primarily the owners of the large tourism boats and the conservationists working for international NGOs. Many fishermen feel that their voices have not been heard and that numerous management decisions regarding the GMR have been influenced mostly by foreign or continental tour operators. As we have seen, in the last decade, the economic importance of fisheries for the Galapagos economy has diminished in a significant way. Whereas in 2003, fisheries represented a total income of seven million dollars to the local economy, by 2006, it accounted for only 2.5 million mainly due to the collapse of the sea cucumber fisheries (Hearn 2006; Quiroga 2011). After economic downturn in the sea cucumber and lobster fishing industries, many in the sector started looking for alternatives in order to survive. Increasingly, they began to propose alternatives that will result in their greater involvement in tourism.

Although the level of conflict diminished significantly after 2004, there is still much animosity between fishermen and conservationists. Fishermen from Santa Cruz and San Cristobal often complain about the amount of money conservationists make as they sell the idea of saving the islands and their creatures. They feel that the islands' endemics, in particular Lonesome George and other tortoises, have been used by the conservationists to gain funds and increase their salaries. They complain that money was spent on removing the tortoises during major volcanic eruptions. When we get sick, nobody cares, say the inhabitants of Isabela, but when a natural event such as a volcanic eruption threatens the tortoises, they are removed by helicopters (see also Ospina 2006 for similar statements). Even as recently as May 2012, in an interview on local radio, Eduardo Veliz, a popular and controversial politician who used to represent the islands in the National Congress, complained that when the electric plant in San Cristobal failed, a young child had to undergo surgery using a physician's cell phone as a makeshift lamp, whereas there is an international outcry each time Lonesome George farts. For many of the local inhabitants and politicians, the local people and the fishermen have been criminalized and blamed for all the perils of the islands, many of which are the result of the mistaken policies of conservationists and the tourism industry (Quiroga 2009a, b).

In his thesis, Pablo Ospina (2004) reproduces some comments that the local people have made to him with respect to some of the local species, illustrating the existence of a discourse against conservation and conservationism. For example, when several sea lions were killed in 2003—a killing that many conservationists blamed on one or several fishermen—fishermen counterattacked, saying that the killing was caused by the conservationists to create a need for their presence. I have heard similar accusations that show that the local people mistrust the conservationists and believe they are to be blamed for the destruction of natural capital. Similarly, in interviews that we conducted and similar ones conducted by Burbano (2011), one can see the anger fishermen feel against some of the emblematic animals that are most dear to conservationists, such as sharks, sea lions, and tortoises. This anger derives not only from practical considerations, such as the fact that sea lions and sharks eat the fish that fishermen catch, sink their boats, and—according to many fishermen—due to their overabundance, decrease the availability of fish in the ocean, but also from the fact that these animals are associated with tourism and conservation. The fact that these stories are still being told in the Galapagos shows that there is still a big gap between the two incommensurable ways of valuing and understanding nature and animals. With a more utilitarian vision that values the direct use of resources, the locals residents value animals based on a very pragmatic scheme, while the scientific system derives from Darwinian and conservationists constructs which are distant and still ungraspable by sectors of the local population. The differences between fishing and conservation illustrate the gaps that exist in other areas of society like agriculture, construction, and other economic activities; these other areas often reproduce the same anticonservationist discourses, as they feel that the excessive controls imposed on them by the GNP and the NGOs are not helping nature but the interests of special groups.

The differences between the two visions were heightened when the dominant conservation and scientific views of the island were operationalized in a series of legal and management schemes. The creation of the protected terrestrial and marine areas in the second half of the twentieth century polarized the two perspectives and accentuated the divisions. The criminalization of many activities that were considered a threat to biodiversity conservation, such as fishing practices and types of gear, of agricultural production techniques, and of construction materials like cutting native woods, was also a process of imposing the scientific evolutionary paradigm of valuing nature, at the expense of the local view. During the 1990s and the early 2000s, when the two groups and their visions were very polarized, commensurability seemed to be a distant possibility, and the most people thought possible was a peaceful coexistence of the groups holding increasingly divergent views.

## A Hybrid Discourse, Land-Based Tourism

The process of expanding paradigms and visions often involves the homogenization of differences and the accommodation of different and diverse interests into unified narrative schemes. This process is often the result of negotiations between actors who control different resources and have different powers. Furthermore, as in the case of the Galapagos, this process is not a one-sided elimination of alternative views but rather the assimilation and accommodation of disparate value systems into new hybrid cosmologies. It also often involves economic transformations and

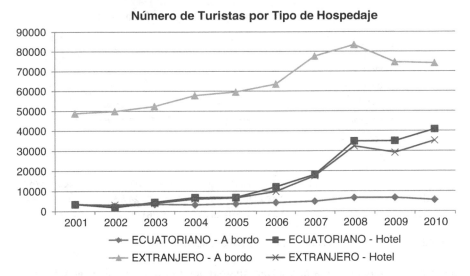

**Fig. 2.1**  Number of tourists per type of accommodation in the Galapagos Islands (Mena 2011)

changes in the material conditions of the different groups. As such, it involves new adaptations to novel ecological, economic, and demographic realities.

Besides the floating hotel operation, another type of tourism has grown in importance during the last three decades. Starting in the 1980s, tourism has been staying increasingly in the towns and using the services of the local population. This type of land-based tourism has become a major part of the economy in all of the islands (Epler 2007) and has been growing in a big way to the point that now almost the same number of visitors goes to hotels and residencies on land and stays on the large cruises. Many of the hotels and residencies are owned by Galapagos residents (Epler 2007). Tourists then travel from island to island on speed boats owned by the local residents and organize daily visits to places close to the ports, often on boats owned and/or operated by fishermen. This type of tourism is attracting mostly local young international tourists and national tourists (tourists from mainland Ecuador), as can be seen in Fig. 2.1 (Mena 2011).

National (Ecuadorian) tourism has increased in recent years and has become a major source of revenue for the local population. It might become as important as international tourism in the near future [see Fig. 2.2 (Mena 2011)].

As they try to increase land-based tourism and reduce the number of tourists staying on cruise boats, local tour agencies, local residents in general, and politicians challenge the dominant discourse and management practices that have been imposed by the conservationists and part of the tourism sector. Local residents and politicians claim that there has to be a change in the exclusive floating hotel model which has been dominant in the past. As I mentioned in a previous article (Quiroga 2009a, b), local people often perceive that there are three sectors: tourism, conservation, and science with the aid of the national government that are seeking to shape and manage the islands for their own benefit, often without considering the needs of the local

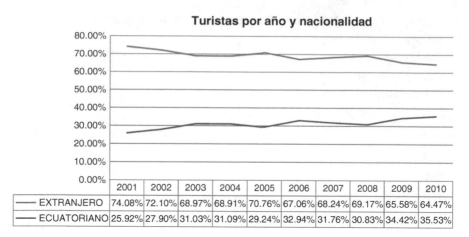

**Turistas por año y nacionalidad**

| | 2001 | 2002 | 2003 | 2004 | 2005 | 2006 | 2007 | 2008 | 2009 | 2010 |
|---|---|---|---|---|---|---|---|---|---|---|
| —— EXTRANJERO | 74.08% | 72.10% | 68.97% | 68.91% | 70.76% | 67.06% | 68.24% | 69.17% | 65.58% | 64.47% |
| —— ECUATORIANO | 25.92% | 27.90% | 31.03% | 31.09% | 29.24% | 32.94% | 31.76% | 30.83% | 34.42% | 35.53% |

**Fig. 2.2** Tourist per year and origin of tourist in the Galapagos Islands

residents. As a response, a new advertising strategy now emphasizes new types of tour packages with homestays and activities based in the towns. A new view, backed in part by some conservationists, is now promoting ecotourism packages. This new strategy is based on new types of activities such as day tours, adventure tourism, *pesca vivencial* (artisanal experiential fishing), catch-and-release sport fishing, kayaking, diving, biking, and sport events like marathons and triathlons, as the types of tourism they feel that will bring real benefit to the communities. There is also an emphasis on defining ecotourism as a type of tourism that benefits not only the environment but also the local population. Local authorities, such as the governor and the mayors, as well as local opinion leaders like radio and TV announcers and businessmen, are trying to promote their towns as tourism destinations, improving boardwalks and building new docks for tourists as part of an effort to attract tourists to their towns. Many young foreign travelers and visitors, including backpackers and large groups of college and high school students and young volunteers, are now staying in the towns. Many homes are offering homestays, and some have even started to build extra rooms to satisfy the growing need for local accommodations.

Besides this type of town-based international tourism, the accelerated growth of national tourism in the last decade has had important implications for the island. Changes in the national economy, such as an increase in oil income, dollarization, and a general increase in the GNP per capita, have meant that the Galapagos is no longer, as it used to be, a destination that only the wealthy and upper classes of the country can afford. Now many, more Ecuadorians from the growing middle class are traveling to the Galapagos on vacation. This new type of tourism uses more of the local facilities, and although Ecuadorian tourists spend less in general than international tourists, more of their money stays within the communities (Taylor et al. 2006; Epler 2007). In an important way, this affects the labor market in the Galapagos as more people are now depending on tourism as their main source of income. Epler (2007) notes that tourism now accounts for more than 50% of the economic activities in the Galapagos, while fishing is only 3%. This new economic reality also manifests itself in the structure of peoples' values, ideas, emotions, and perceptions. Thus, from this new economic reality which includes the collapse of the fisheries and the increasing importance of tourism for the global population, a new discourse is emerging: one that sees the need to conserve the islands' resources but considers that these efforts cannot benefit only outsiders who do not reside in the islands. Many local people no longer see tourism as a foreign and negative force, but rather as something that they need to know how to participate in and from which they can benefit. The original frontier mentality based on extractive industries, fishing, and agriculture is now adapting to this new reality and developing new types of hybrid understandings and sensitivities.

## A New and Emergent Hybrid Culture?

The Mexican author Nestor Garcia Canclini (2001) noted that social scientists have often overlooked the complexities associated with the production of new cultures, failing to examine the manner in which different discourses generate conflict and opposition as well as how negotiation and accommodation generate shared views or hybridizations and, possibly, commensurable visions. As we have seen above, historically two incommensurable discourses dominated the way visitors and residents perceived the Galapagos: a scientific–conservationist globalized view and a local frontier and utilitarian mentality (Quiroga 2009a, b). During the last part of the twentieth century, these two views became more polarized, and conflicts emerged mainly as the result of the fight between fishermen and conservationists. With the creation of the Charles Darwin Foundation and the GNP, institutional support was generated for the conservationist discourse. As experts and professionals entered the scene, conservation was problematized, and a discourse was produced about the need to protect the Galapagos, in particular from the local population, for the rest of humanity. UNESCO played an important role in establishing the discourse. The global position has been effectively imposed over the local view classifying the local activities as more or less adequate and criminalizing many of the behaviors of the local residents and producing a value system in agreement with the Darwinian paradigm. For a period of time, that meant that the two systems coexisted in heterodoxia, without much dialog between them.

During the last part of the twentieth century, a new hybrid discourse was created based on the traditional framework of the local residents and the assimilation of many conceptual schemes and sensitivities from the discourse of conservationist and tourism operators and the conservationists, changing their strategies to include the local people in their conservation agenda. This new discourse was composed of many bridging concepts that were developed as each of the systems adapted and accommodated to the others and each of the views assimilated aspects of the others in a process of negotiation between different actors. In general, one can say that the local system having had less access to resources had to incorporate more elements from the global view of the islands. As we have also noticed, economic transformations caused especially by the increase in the number of tourists staying on land have played a key role in the cultural transformations that are occurring in the islands.

During the 1990s, especially during the late 1990s, the Charles Darwin Foundation started to incorporate some of the new social reality as part of its discourse to include the local population in the planning and execution of different conservation programs. During the late 1980s and especially the 1990s, some of the key producers of the traditional discourse of conservation, which negated the role of the local people such as the Charles Darwin Foundation and other NGOs, saw a need to change their strategy and started to talk about education and incorporating views and perspectives of the local population. The new conservationist model needed now to include the existence of a local population which could no longer be ignored. As conservation is being reframed as a social problem needing social science expertise, new professionals and organizations have begun to enter the scene. Educational and public awareness campaigns, as well as a changing economic reality, are now transforming the traditional local framework.

In general, most of the population has assimilated many of the constructs and sensibilities of the global environmental discourse to different degrees and with various levels of sincerity. There are, however, still important sectors that maintain a more traditional framework. Many of these more traditional constructs of nature that reflect, to a large degree, the frontier mentality discussed above can be found in the rural areas and among the fishermen and the agriculturalists. In a survey conducted in 2009, we interviewed 210 residents in San Cristobal about the position of the people with respect to sea lions. It became clear that the community is divided with respect to the value of these animals. For 66% of those associated with the fishing sector, sea lions are more a menace and constitute a problem when they are trying to do their jobs. However, for most of the residents interviewed, the animals represent a symbol of their town and are viewed as important because they attract tourists; thus, most people in San Cristobal perceived the animals in a positive way. A large number of residents—69% in the community in general and 66% of those involved in tourism—believe that the sea lions are cute and fun, whereas only 28% of the fishermen felt that way. Sea lions in the Galapagos are viewed as part of the Galapagos ecological identity and also as an economic counter force to local fisheries. These alternative perspectives represent the complexities of accommodating multiple visions in the Galapagos. Galapagos residents seem to be living a moment of transition where a new understanding is emerging from the polarized past. This new hybrid view takes important elements from the traditional science, conservation, and tourism discourse while maintaining the idea that the local residents have a right to use the rich natural resources of the islands and to shape and be shaped by the social–ecological interactions that define the Galapagos.

From this and other similar interviews, we can conclude that for some sectors of society, changes in the way people value their resources occur slowly and, in the case of some sectors, like agriculture and parts of the fishing sector, it has affected them little. Thus, often the value system and the sensitivities of fishermen who are now working on tourism are, to a large extent, the same as they were before they changed their activities. When they feel that the GNP is trying to stop them from profiting from tourism, they still mention as a threat that if they are not allowed to profit from tourism, they will go back to shark finning.

For most of the population, however, a new type of environmentalism is emerging, one that is closer to what Martínez Alier (2007) has termed popular environmentalism, an environmentalism that is based on the preservation of natural places and biodiversity not for its own sake but for the benefit of the people, especially poor people, living next to these resources. A series of anecdotes and stories illustrate the change in strategy and practice. As we were going with a group of students to Kicker Rock on a day trip, the captain of the boat in Isabela told me how fishermen on that island are now more interested in taking tourists snorkeling than in going fishing and are buying live animals from other fishermen, such as sea horses and octopus, and taking them to places where they later will take the tourists. A dive master who used to be a fisherman told me with sadness how he used to kill sharks, but now that he is diving with tourists, he sees how beautiful they are under the water. Another fisherman who used to kill sharks admitted that he has now stopped doing so because his kids complained each times he arrived home with shark fins. This new hybrid discourse is thus starting to question some of the old dichotomies between the global and the local and conservationism versus extraction.

Both governmental and nongovernmental organizations such as the Araucaria Project (Spanish cooperation), the Charles Darwin Foundation and the World Wildlife Fund as well as private companies like SCUBA Iguana have trained fishermen to become diving guides. Although many of these efforts have not been successful (for few of them are actually working as SCUBA diving guides), some have converted to working in tourism, and there are now some fishermen who get most of their income from tourism and guiding. Also the fishermen and other local residents have produced different projects involving a change to, what they argue, more sustainable activities. Probably one of the most controversial is that of *pesca vicencial*. The basic concept is that fishermen take tourists with them for a day of fishing in the traditional way. The justification is that in this way, they will be decreasing their fishing efforts, thus, the extraction of resources. The idea was originally proposed by fishermen like Carlos Ricaurte of San Cristobal. NGOs and the GNP have supported the efforts of several fishermen. At the moment, according to the GNP web page, there are 24 boats belonging to Galapagos fishermen who have a permit to do *pesca vivencial*. For some fishermen, *pesca vivencial* is not a viable alternative, and they have proposed instead to do sport fishing catch and release, for they argue that sport fishing that targets large fish such as bill fish will attract more international attention. This strategy is something that the GNP and the CDF have questioned, and they have said that they oppose the idea of sport fishing as a tourism alternative in the Galapagos. The popularity of these new and often controversial ideas does not necessarily mean that the fishermen have shifted completely to a new value structure; rather, one must see them as making a strategic move as they try to access new types of resources and learn how to negotiate within the spaces left open by the dominant discourses and practices.

Since 2004, the tensions between fishermen and conservationists started to decrease as a result of several events, such as the diminished importance of fishing for the Galapagos economy; the increased interest in tourism by many permanent residents, including fishermen; changes at the national level as some political parties disappeared from the scene; and a change in conservationist discourses and practices toward becoming more aware of the need to include local people in their strategies. Conservationists started to perceive local inhabitants as a necessary part of their strategy to save the islands. The facts that fishermen now perceive that their income might be threatened by problems encountered by some of the major fisheries have meant that for many of them, tourism is the only realistic alternative. This new situation has become an important factor in shaping the attitudes of fishermen, especially the young ones, vis-a-vis tourism, and conservation. In her interviews with fishermen for her MA thesis, Diana Burbano documented not only the fact that many fishermen already have started to get involved in tourism but also that many of the young (49%) and the middle-aged (35%) fishermen would like to see tourism rather than fishing as their main activity because they make more money from it and it is less demanding (Burbano 2011). As new practices and economic systems emerge, such as sport fishing, day tours, SCUBA diving, surfing, and kayaking, the gap between the global and the local discourse has narrowed. This new emergent conceptual system builds on the rejection of the traditional conservation and tourism models, which many local people consider have failed in protecting the islands' resources and improving the well-being of the people, while at the same time, it appropriates some key concepts and symbols from scientific and conservationist cosmology, like the importance of conserving endemic and native animals and plants. No longer can we say that most of the population of the Galapagos perceives conservation as a dominant and external strategy; rather, there is now a sense that much of their well-being depends on their successful management of natural resources.

Tourism, which is now the main economic engine in all of the islands, has had a tremendous influence on the economy and on people's livelihoods. In Isabela, there are now some 20 fishermen who work at the dock doing bay tours and taking tourists to visit the *poza de las tintoreras*, the other side of the bay. Most of them have practically stopped fishing. These fishermen have invested in improving the level of comfort of their fiber glass fishing boats to accommodate the tourists. As is also the case in San Cristobal, most want to dedicate more resources and time to tourism, which they feel is less demanding and more profitable. Several claim that the number of people doing shark finning has decreased significantly because they now have an alternative, but they threaten to go back to their original activity if the park is going to regulate their activities. Similar processes are occurring in all the islands as fishermen are working on different boats as captains and sailors. In other words, most fishermen are very pragmatic about their greater acceptance of the conservation perspective. They feel that as long as it is convenient for them to conserve, they will do so, but once that is not the case, they will go back to their old practices.

Most of the local residents are now in agreement with the general principles of conservation, as can be seen from the results of Instituto Nacional de Estadistica y Censos (INEC). When asked if the resources must be conserved in the long run, 75.1% of the population of the Galapagos answered yes. However, it is fair to say that it is among the younger population where the change is more evident. In many of the workshops and classes that I have conducted with high school students and local college students in San Cristobal, it is clear that many of the younger people have a much more sincere commitment to the principles of conservation. Many in this age group (between 15 and 22 years old) feel that conservation is an imperative and they have a real responsibility for the natural world. Words like climate change, sustainable energy, and waste recycling are now becoming part of young peoples' everyday discourse. They often complain about the attitudes of their elders, who they feel do not understand the importance of resource conservation. This generation gap that has been created is partly the result of education campaigns that NGOs, the GNP, municipalities, and universities have been promoting.

This change in attitudes is in part due to the resources available to conservationists to spread their message. There are now numerous programs to increase the awareness of the local population about conservation. The GNP has a radio and TV program which talks about the achievements of the park; the Charles Darwin Research Station has different educational initiatives and centers to run campaigns. Universities like the Universidad San Francisco de Quito (USFQ) and Universidad Central have also created majors such as natural resource management and ecotourism for the local population that include the teaching of conservation and evolution.

This new hybrid culture, rather than rejecting science and conservation, demands that scientists ask new types of questions. Quiroga and Ospina (2009) conducted a series of interviews regarding the acceptance of science among the local people. We found that 84.2% of the population thought that more scientific research was needed. Also, a large portion of the local people—when asked about the role science must play—said that science should be involved in studying the impacts of migration, in public health, and in the impacts of tourism (Quiroga and Ospina 2009). From this survey, we can say that a large section of the people of the Galapagos now views science as a potentially beneficial institution, but considering that rather than emphasizing the study of evolution and other traditional biological and geological issues, science must be directed to solve the problems and issues that affect people.

Even the most cherished symbols of science are being integrated into new hybrid constructs. Darwin's name and image are now used by the local population in many and often creative ways. His image has been shaped and transformed according to the needs and perspectives of the local population. The large towns now have streets and plazas named after him, and several public places and buildings carry his name. This is the case in San Cristobal, where the municipality has named the newly remodeled conference center the Charles Darwin Convention Center and has placed a bust of Charles Darwin on the boardwalk and his statue in Tijeretas, the place where he first landed in 1835. Despite the growing presence of religious groups, such as different Catholic groups, Jehovah's Witnesses, Mormons, and Seventh Day Adventists, among the residents, the image of Darwin has been accepted, appropriated, and used by the local population. In the same way that nature has been stripped of its most brutal and discomforting aspects in its presentation for tourists, so have Darwin and Darwinism been stripped of their most secular and bothersome interpretations by the residents of the Galapagos. Darwin's image is, thus, no longer just the icon of the international scientific community and the global conservationist discourse but also, in a transformed and adapted manner, an icon for the local population.

## Isolation or Connectivity: The Framing of a New Problem

As the discourse that the main conservation problem of the Galapagos is the fight between extractive and non-extractive activities loses its relevance, other forms of problematizing conservation are regaining more importance. Isolation has been seen as a key concern for scientists and environmentalists since the middle of the twentieth century. Both biophysical and socioeconomic factors have affected the high degree of isolation which characterizes the islands. Grenier (2012) has described the socioeconomic threat as the continentalization of the islands (i.e., the islands becoming more like the mainland). As he has indicated, there are both national (Ecuadorian) and international factors that have influenced the pace and degree of the connectivity between the islands and the mainland and between islands. These factors, which include the oil boom that the country experienced in the 1970s, the dollarization of the economy, and the international demand for products such as sea cucumber, shark fins, and lobster, have all steadily increased the connectivity of the islands with the mainland. He has also noted how the degree of connectivity has continue to increase, despite the efforts of the 1998 Special Law and the creation of the Galapagos Marine Reserve to curb the increase in connectivity by applying strategies such as limiting immigration, industrial development, and the expansion of the fisheries.

Tourism, migration, and the increased importation of goods are seen as some of the most important threats to isolation. Increased connectivity threatens the natural laboratory, as it can cause changes to habitats, it threatens animals and plants directly with invasives and can result in the mixing of species that have developed in isolation. Some people think that the fight has already been lost. A very controversial article by a scientist (Gardener 2011) suggests that it is time to learn how to live with invasive species. Questioning the duality between isolation and connectivity, the argument is a direct criticism of all the multimillion dollar eradication campaigns, many of which have not worked. With new discourse about the loss of isolation, there are increasing debates about the best way to manage invasives and population growth and to increase the number of tourists. All of these are part of problematizing the Galapagos by NGOs, scientists, journalists, and government officials. This problematization has underscored many incompatibilities between the global and local discourses.

As in the case of the previous discourse against the extraction of natural resources, local people are starting to respond to the idea of the need for greater isolation and less consumption. The way local discourses have been dealing with this new problem is through the concept of *tranquilidad*. The anthropologist Pablo

Ospina has noticed the central cultural importance of this concept for the local culture (Ospina 2006). People in San Cristobal complain about the fast pace of life in Santa Cruz and the way in which consumerism has come to dominate the island mentality. The elders often talk about the past as an ideal time when the stress and tensions of modern life were not as pervasive as they are today. A rejection of the speeding up of the pace of life is now seen by some locals as an important base for maintaining a more sustainable relationship with nature and others. This rejection is, as I was told by some of my local students a life choice, *una opcion de vida*, a more sustainable alternative, which they feel needs to be valued by all. Many of the inhabitants of San Cristobal, at least at the level of discourse, have rejected consumerism and modernity. This idealized version of Galapaqueno culture and values contrasts with the increasing number of cars, scooters, air conditioners, household appliances, computers, and other amenities that are finding their way to the islands. In reality, the gap is growing again between the local and the global discourses. This time, it is between a view that sees the value of the islands as a natural laboratory and the residents' view of the islands as a place where they can make a comfortable living. As the population increases and becomes more affluent, there are now concerns about the increased consumption of people living in the islands.

The standard of living in the Galapagos is relatively high compared to the rest of the country. The dream some local people have of the Galapagos as an isolated and tranquil place contrasts with an ever-increasing degree of continentalization driven by an even more powerful desire to be connected to the mainland. The increased number of tourists is also associated with more frequent flights from the mainland. Rising numbers of ships coming from the mainland with cargo and increases in the access to goods and services are part of the new Galapagos. At the moment, there are 12 commercial flights arriving in San Cristobal each week and 31 arriving in Santa Cruz (Freddy Valenzuela, San Cristobal Airport, personal communication). The number of goods brought from the mainland is also increasing. Most of the residents have many home appliances. According to INEC, in 2009, 93.9% of the families had color TVs, 92.5% had cellular phones, 44.3% had a computer, and 30.3% had access to cable TV (INEC-CGRE Encuesta de Condiciones de Vida 2009). This demonstrates not only the need for products and goods coming from the outside but also how well connected the local population is to the rest of the country. This trend for greater connectivity is also seen in the case of the desire of the local population for more and easier means of transporting people and goods from the mainland and between islands. Thus, according to INEC, 83.1% of the people would like the number of flights to and from the mainland to increase, 67.3% would like to see more air transportation between islands, and 64.6% would like marine transportation with the mainland to increase. In 2012, when the government tried to better regulate marine transportation and forbid some boats from coming to the islands as they did not comply with the conditions established by the GNP, the local population complained bitterly about the lack of access to imported goods. They also responded positively to having more land-based tourism (63.3%), much higher than the number who would like to see cruise boat tourism increase (43.3%) (INEC-CGRE Encuesta de Condiciones de Vida 2009).

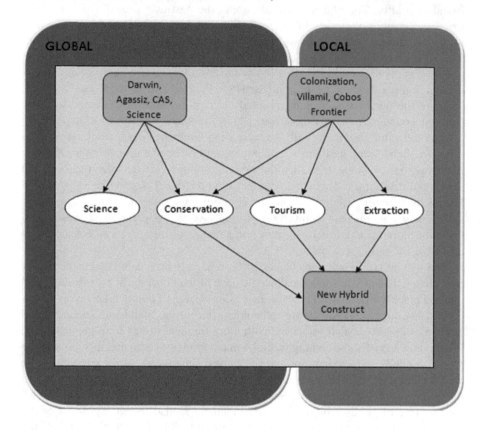

## Conclusion

The Galapagos' status as a natural laboratory, the conservation problems facing the islands, and possible solutions have been defined and framed to a large extent by international bureaucracies, NGOs, and Ecuador's national government. This global view has been confronted and challenged by local definitions which were the result of the process of colonization and the opening of a frontier by different groups of Ecuadorian pioneers. As we have seen, in a series of transformations, views about the islands have shifted from being divergent and incommensurable to becoming hybrid products of negotiations and impositions. However, as solutions to the old conflicts are discovered, new ones arise that must find a negotiated consensus.

Charles Darwin, who was to a large extent responsible for the greatest paradigm shift in biology, also played a critical role in shaping the vision and understanding the world has of the Galapagos. As he visited the Galapagos and later thought and wrote about his discoveries on the islands, he established their importance as a natural laboratory and defined a place that could act to solve future scientific debates that blossomed from his theory. This idea, which is based on several biogeographical characteristics of the islands, motivated a series of expeditions, visits, and studies by renowned scientists, many of whom wanted to question or revise Darwin's original observations and conclusions. In the twentieth century, a marriage was created between conservation and evolutionary science in the Galapagos and other parts of the world, which has been so successful that today we have come to think of this relationship as a natural association. Based on the global scientific–conservation vision of the Galapagos comes the idea explored and utilized by the floating hotel tourism sector that marketed the idea of the Galapagos as a pristine natural paradise.

A few years before Darwin arrived, Ecuador had claimed possession of the islands and had sent groups of colonists to assure control of the territory. These pioneers transformed the Galapagos and reproduced distant settings from which natural resources could be extracted, a process and a view that are incommensurable with the scientific construct of the Galapagos Archipelago. The view of the Galapagos as a frontier necessitated that it be conquered, subdued, civilized, and domesticated, a subjugation of nature played out against a constant struggle for supremacy. As these colonists shaped the Galapagos according to their own perceptions, needs, and expectations, they threatened the isolation required by many of the key biological process that fed the scientists and their paradigm. As one can expect, conflicts between the sectors were inevitable, and violent clashes between the groups erupted as the diverse interests and discourses met.

Although one can say that the two views are separated by a wide conceptual and economic gap, among some sectors, the distance between these diverse and opposing views has slowly become narrower as both conservationists and scientists now consider it impossible to conserve the islands without the support of local people. Local people are starting to realize that if they are to benefit from the natural resources of the islands, they need to use long-term conservation strategies. A new hybrid view has developed by which elements of the different discourses are mixed, creating new mosaic visions. There has also been an important change in the way conservationists face social issues. Since the last part of the twentieth century, they have incorporated local people in most of their strategies and have included social scientists and professionals. Thus, incommensurable views and perspectives are finding bridges and points of encounter, the result of constant negotiations that seem to narrow the gap between the two original perspectives. Tourism and ecotourism and the growing access to material and cultural resources that this industry brings are one of these points of contact between the otherwise divergent perspectives. It is a vital issue for the Galapagos how these different visions and sensibilities will transform and shape each other in the future. The extractive activities are no longer seen as the major challenge for the islands; the new challenge is now problematized as the increasing degrees of connectivity and continentalization of the local population. As this challenge becomes more relevant, new solutions must emerge and be negotiated.

Interviews were conducted in San Cristobal with 12 fishermen and in Isabela with four fishermen. I also interviewed local students from the university and participated in a workshop organized by the Charles Darwin Research Station in Isabela. Interviews in the highlands of San Cristobal were also conducted, and class discussions and conversations with local students at GAIAS and workshops with high school students were held about the environment, science, and climate change. Lastly, discussions and meetings with authorities and other leaders were also held to create this contemporary view of the Galapagos.

# References

Agassiz L (1896) A journey in Brazil. Houghton, Mifflin, and Company, Boston
Baur G (1891) On the Origin of the Galapagos Islands. Am Nat 25(292):307–326
Becken S (2010) The importance of climate and weather for tourism. Lincoln University, Lincoln, NZ
Bourdieu P (1984) Distinction: a social critique of the judgment of taste. Harvard University Press, Cambridge, MA
Burbano D (2011) Shifting baseline en la pesca blanca de Galapagos: relaciones socio-ecológicas en ambientes marinos. M.A. thesis, Universidad San Francisco de Quito
Cairns R (2011) A critical analysis of the discourses of conservation and the science of the Galapagos Islands. Ph.D. dissertation, University of Leeds, UK
Castrejón M (2006) El sistema de co-manejo pesquero de la reserva marina de Galapagos: tendencias, retos y perspectivas de cambio. Fundación Charles Darwin, Santa Cruz, Ecuador
Dexter RW (1979) The impact of evolutionary theories on the Salem Group of Agassiz zoologists (Morse, Hyatt, Packard, Putnam). Essex Inst Hist Collect 115(3):144–171
Durham W (2012) What Darwin found convincing in Galapagos. Routledge Explorations in Environment Economics. The Role of Science for Conservation, New York, pp 3–15
Epler B (2007) Tourism, the economy, population growth and conservation in Galapagos. Document Submitted to the Charles Darwin Research Station, Puerto Ayora, Ecuador
Escobar A (1994) Encountering development: the making and unmaking of the Third World Princeton. Princeton University Press, NJ, pp 1–320
Espeland W, Mitchell S (1998) Commensuration as a social process. Annu Rev Sociol 24:313–343
García Canclini N (2001) Culturas Híbridas: Estrategias para entrar y salir de la modernidad. Estado y Sociedad, Paidós, p 352
Gardener M (2011) Seeking novel solutions to manage native biodiversity. Conservation Ecology. Charles Darwin Foundation, Santa Cruz, Ecuador
González J, Montes C, Rodríguez J, Tapia W (2008) Rethinking the Galapagos Islands as a complex social-ecological system: implications for conservation and management. Ecol Soc 13:13
Grant PGR (2008) How and why species multiply. The radiation of Darwin's finches, vol 95. Princeton University Press, Princeton, NJ
Grenier C (2007) Conservación Contra Natura. Abya Yala, Quito
Grenier C (2012) Nature and the world: a geohistory of Galapagos. In: Wolff, M. and Gardner, M. (eds) The role of science for the conservation of the Galápagos: a 50 years experience and challenges for the future. Routledge (London) pp 256–274
Hearn A (2006) Evaluación de la Pesquería de Langosta Espinosa (*Panulirus Penicillatus*) en Evaluación de la Pesquería en la Reserva Marina de Galapagos. 46

Hearn A (2008) The rocky path to sustainable fisheries management and conservation in the Galapagos Marine Reserve. Ocean Coast Manage 51(8–9):567–574. doi:10.1016/j. ocecoaman.2008.06.009

Hennessy E, McCleary A (2011) Nature's Eden? The production of pristine nature in the Galapagos Islands. Isl Stud J 6(2):131–156

Howes H (2011) Oceanic encounters: exchange, desire, violence. J Pac Hist 46:136–138. doi:10.1 080/00223344.2011.573645, Retrieved from http://dx.doi.org/

James M (2010) Collecting evolution: the vindication of Charles Darwin by the 1905–1906 Galapagos expedition of the California Academy of Sciences. In: Proceedings of the California academy of sciences series 4. San Francisco, California: 61:Sup II, No. 12, pp 197–210, 3 fig

Larson E (2001) Evolution's workshop: God, science on the Galapagos. Penguin Books, London

Latorre O (2001) La Maldición de la Tortuga: Historias Trágicas de las Islas Galapagos (Cuarta Edi). Artes Gráficas Senal, Impresenal Cia, Quito, Ecuador, p 226

Martínez Alier J (2007) El ecologismo popular. Ecosistemas 16(3):148–151

Mena C (2011) Tendencias del Turismo de Galapagos en la Ultima Década. San Cristobal, Ecuador

Morris HM (1988) Men of science, men of God: great scientists who believed the Bible. Master Books, Green Forest, AR, p 56

Obeyesekere G (1997) The apotheosis of Captain Cook: European mythmaking in the Pacific. Princeton University Press, Princeton, 71, p 313

Orduna J (2008) Ecofacismo: las internacionales ecologistas y las soberanias nacionales. Martinez Roca, Buenos Aires

Ospina P (2001) Identidades en Galapagos: el sentimiento de una diferencia, pp 1–90

Ospina P (2004) Galapagos, naturaleza y sociedad: actores sociales y conflictos ambientales. Universidad Iberoamericana, Disertación

Ospina P (2006) Galapagos, naturaleza y sociedad: actores sociales y conflictos ambientales. Corporación Editora Nacional, Quito

Povinelli E (2001) Radical worlds: the anthropology of incommensurability and inconceivability. Annu Rev Anthropol 30:319–334

Proctor J (1998) The social construction of nature: relativist accusations, pragmatist and critical realist responses. Ann Assoc Am Geogr 88:352–376, http://www.jstor.org/stable/2564234

Quiroga D (2009a) Galapagos, laboratorio natural de la evolución: una aproximación histórica. In: Tapia et al (eds) Ciencia para la sostenibilidad en Galapagos. Parque Nacional Galapagos, Puerto Ayora, Galapagos, pp 13–61

Quiroga D (2009b) The Galapagos: the crafting of a natural laboratory. J Polit Ecol 16:136

Quiroga D (2011) Biodiversidad de Galapagos, Amenazas y Oportunidades para la Conservación. Manual de Aplicación del Derecho Penal Ambiental como Instrumento de la Proteccion de las Áreas Naturales en Galapagos. Sea Shepherds, World Wildlife Fund, and Galapagos Academic Institute for the Arts and the Sciences, Universidad San Francisco de Quito, Quito, Ecuador

Quiroga D, Ospina P (2009) Percepciones sociales sobre la ciencia y los científicos en Galapagos. In: Tapia et al (eds) Ciencia para la sostenibilidad de Galapagos. Parque Nacional Galapagos, Puerto Ayora, Galapagos, pp 109–126

Quiroga D, Mena C, Bunce L, Suzuki H, Guevara A, Murillo JC (2010) Dealing with climate change in the Galapagos: adaptability of the tourism and fishing sectors. In: Larrea I, Di Carlo G (eds) Climate change vulnerability assessment of the Galapagos Islands. World Wildlife Fund and Conservation International, USA

Sahlins M (1995) How "Natives" think: about Captain Cook, for example. University of Chicago Press, Chicago, p 328

Sulloway FJ (1982) Darwin and his finches: the evolution of a legend. J Hist Biol 15(1):1–53

Sulloway FJ (1984) Darwin and the Galapagos. Biol J Linn Soc 21:29–59

Taylor JE, Hardner J, Stewart M (2006) Ecotourism and economic growth in the Galapagos: an island economy-wide analysis. Environ Dev Econ 14:139–162

Tcherkezoff S (2009) A reconsideration of the role of Polynesian women in early encounters with Europeans: supplement to Marshall Sahlins' voyage around the islands of history. In: Jolly M et al (eds) Oceanic encounters: exchange, desire, violence. ANU ePress, Canberra Australia, pp 113–159

Valle C, Parker P (2012) Research on evolutionary principles in Galapagos: an overview of the past 50 years. Routledge Exploration in Environment Economics, The Role of Science for Conservation, New York

Watkins G, Cruz F (2007) Galapagos en riesgo: un análisis socioeconómico de la situación actual en el archipiélago. Fundación Charles Darwin, Puerto Ayora, Galapagos

Winsor MP (1979) Louis Agassiz and the species question. Stud Hist Biol 3:89–138

# Chapter 3
# Perspectives for the Study of the Galapagos Islands: Complex Systems and Human–Environment Interactions

Stephen J. Walsh and Carlos F. Mena

## Introduction

"Enchanted Islands," "Ecological Paradise," "In the Footsteps of Darwin"—we have been intrigued by these, and many other, colorful descriptions of the Galapagos Islands and have enjoyed the many declarations of the islands' mysterious uniqueness, well proclaimed by ancient mariners, explorers, pirates, and scholars. Their accounts are testimony to the presence of the "Imps of Darkness" and the "Fire in the Earth," as new species and new landforms combine to create a landscape and seascape populated by endemic inhabitants who have evolved and adapted in very remarkable ways in response to a dynamic environment. In recent times, the hushed mention of "Paradise Lost" and "Paving of Paradise" has become part of the contemporary story of the Galapagos Islands, a story often muted by the amazing descriptions of the iconic and emblematic species that help define the islands and add to their mystery. But with all the splendor and majesty, the Galapagos Islands are in crisis; a crisis born of the very success that the archipelago has enjoyed in maintaining its native and endemic species at which the world marvels. Historically protected through geographic isolation, the islands are no longer remote; they are now explicitly connected to the global economy and to international tourism markets. Modern travel has effectively created a "land bridge" that connects the

S.J. Walsh (✉)
Department of Geography, Center for Galapagos Studies, Galapagos Science Center,
University of North Carolina at Chapel Hill, Chapel Hill, NC, USA
e-mail: swalsh@email.unc.edu

C.F. Mena
College of Biological and Environmental Sciences, Galapagos Science Center,
University of San Francisco, Quito, Ecuador
e-mail: cmena@usfq.edu.ec

S.J. Walsh and C.F. Mena (eds.), *Science and Conservation in the Galapagos Islands:*     49
*Frameworks & Perspectives*, Social and Ecological Interactions in the Galapagos Islands 1,
DOI 10.1007/978-1-4614-5794-7_3, © Springer Science+Business Media, LLC 2013

Galapagos to the world community, with all of its consumptive demands, threats of invasive species, and tourists craving additional services and richer experiences, generally accommodated by an influx of people mostly arriving from mainland Ecuador, who seek better jobs and improved economic opportunities in the tourism industry. Eager to visit, experience, and/or work in the Galapagos Islands, the residential and tourism populations have substantially increased over the past 20 years, approaching 30,000 residents and 185,000 visitors as of 2011.

Today, the Galapagos Islands must rely upon adaptive and participatory management and enlightened public and environment policy to ensure their survival. The direct and indirect consequences of the expanding human imprint have signaled a concern about the future of the Galapagos. In this interconnected world, the "Galapagos Paradox" will surely be tested—how can the Galapagos Islands be protected from the many endogenous factors and exogenous forces that shape, reshape, and often debilitate many island ecosystems? Can the Galapagos Islands accommodate the increasing levels of tourism, local development, and population migration associated with new residents and international visitors? Drawn by the many attractions of this special place, some are seeking economic rewards, while others seek rewarding ecological experiences and the promise of an ecosystem at peace with its surroundings. The very features and specialness of the Galapagos that attract visitors from around the world and create employment opportunities, mostly in tourism for a migrant population, are the same forces that put stress on vulnerable settings and make island sustainability often just a dream. The challenge is to create a comprehensive and adaptive model of the Galapagos that effectively integrates people and the environment within a complex and dynamic system, where critical thresholds, feedback mechanisms, and nonlinear relationships are recognized within the context of social–ecological dynamics and the factors that induce changes in system behaviors and chart alternate trajectories of the future. We term the interactions between people and environment and their link effects *Island Biocomplexity*, a new framework and perspective for the study of social and ecological systems in the Galapagos Islands and other similarly challenged island ecosystems around the globe.

## Study Area

The Galapagos Islands are a "living laboratory" for the study of evolution, environmental change, and conflicts between nature and society. Free from human predators for almost all of their history, these islands have developed some of the most unique life forms on the planet, adapted to their harsh surroundings and living in ecological isolation. It was not until Charles Darwin's famous visit in 1835—which helped inspire the theory of evolution—that the islands began to receive international recognition. The Galapagos Archipelago encompasses 11 large and 200 small islands totaling approximately 8,010 km$^2$.

In 1959, the Galapagos National Park (GNP) was created, and in 1973, the archipelago was incorporated as the twenty-second province of Ecuador. UNESCO designated the Galapagos as a World Heritage Site in 1978. The islands were further deemed a Biosphere Reserve in 1987. In 1998, the Ecuadorian government enacted special legislation for Galapagos in an effort to promote both conservation of terrestrial and marine biodiversity and sustainable development. The Special Law for Galapagos characterizes introduced species as the principal obstacle to the aim of harmonious coexistence between people and the unique flora and fauna of Galapagos. The Special Law is now being revised as a consequence of the new Ecuadorian constitution that was approved in 2008 through a national referendum.

The Galapagos National Park comprises 97 % of the land areas of the archipelago. The remaining 3 % includes urban areas and agricultural zones. The Special Law implemented a registration system in 1998 to monitor the existing human population in the islands. A more rigorous registration system that tracks movement in and out of the Galapagos is now being implemented. Currently, the Special Law defines four types of people: (1) undocumented or "illegal" workers from the Ecuadorian mainland, (2) "permanent residents," (3) "temporary residents" or workers subject to legal residence restrictions of labor contracts, and (4) "tourists."

During the past three decades, dramatic social–ecological changes have threatened the social, terrestrial, and marine ecosystems of the Galapagos. Beginning in the 1970s, the islands started to experience exponential population growth. Thousands of new residents began to migrate from the mainland, attracted by the promise of lucrative opportunities linked to the islands' rich marine and terrestrial ecosystems and "pushed" by the lack of economic opportunities in many parts of mainland Ecuador. The local population grew from under 10,000 residents in 1990 to nearly 30,000 in 2011. In addition to settlement and population in-migration, the number of tourists has increased from about 41,000 in 1990 to nearly 185,000 in 2011. Some of the more pronounced trends associated with increased human presence include (a) unprecedented use and extraction of terrestrial and marine resources, (b) introduction and proliferation of invasive flora and fauna that can replace native and endemic species, (c) increased degradation of fragile environments, (d) unprecedented energy consumption and waste generation associated with population and tourism growth, and (e) increased interinstitutional conflicts over governance and policy.

## The Galapagos as a Socio-Environmental System

The links between people and environment in the Galapagos serve to frame the many conflicts between and among the various resource conservation and economic development sectors that often have competing interests. Historically, there have always been sectors of the local economy supported by agriculture, as well as by the fishing and tourism industries, but the rapid increase in the economic drivers associated with fisheries and tourism over the last 20 years has exacerbated an already difficult and complex situation. For instance, economic diversification from agriculture to tourism has led to labor shortages on the farm and a demand for mainland immigrants, and a decline in management of invasive species has led to land abandonment and the threat of invasive species "escape" from human use zones to the Galapagos National Park.

The primary decisions of concern at the individual and household levels are related to alternative strategies of household livelihoods and feedbacks to the intensification or abandonment of agricultural activities in response to changing economic opportunities. An important landscape dynamic is the arrival and spread of invasive species. The primary exogenous influences include the changes in the intensity and frequency of El Niño events and the growth of tourism in the islands and related consequences for alternative household livelihood strategies. As tourism increases, it alters the economic and demographic processes at the household level that subsequently affect the way that households manage the landscape.

The ecological system is the focal point of international interest in the Galapagos. Among the greatest threats to the ecosystem of the Galapagos is the growing number of exotic plant and animal species (Mauchamp 1997; Tye et al. 2002; Tye 2006). Increased human presence has hastened the introduction of invasive species that are now so prevalent that they threaten the native and endemic flora and fauna of the islands and significantly impact the human population. The problem of invasive species illustrates the important feedbacks between the social and ecological systems: land management practices reflect human migration patterns and economic choices, whereby increasing urbanization is linked to tourism and other opportunities that render lands underutilized and abandoned, becoming fertile ground for invasive species. To the fragile ecosystem of the Galapagos, these invasive plants change the biological diversity, degrade ecological services, reduce the number of endemic plants, change grasslands to forests, modify ecological processes, and compete with other species.

## Complexity Theory and Agent-Based Models

Complexity theory sees the complex nature of systems as emerging from nonlinearities due to interactions involving feedbacks occurring at lower levels of social and ecological organization within the system (Cilliers 1998; Malanson 1999; Matthews et al. 1999; Manson 2001). Complexity draws on theories and practices from across the social, natural, and spatial sciences (Parker et al. 2003). Of particular interest have been the characterization of spatial patterns and links to processes, feedback mechanisms and system dynamics, and space-time lags and scale dependencies of processes and actors (Evans and Kelley 2004; Malanson et al. 2006). Complexity science offers a new science epistemology focusing on the creation of order by self-organizing heterogeneous agents and agent-based models. The fundamental element of complex systems is the adaptive behavior of human–natural environments (Parker et al. 2003) and how agents learn, react to new conditions, alter relationships with a changing environment, and mediate their behavior relative to external forces, such as climate change and public policy, and endogenous factors, such as the spread or eradication of invasive species.

Agent-based models (ABMs) are constructed in a "bottom-up" manner by defining the model in terms of entities and dynamics at a micro-level, that is, at the level of individual actors and their interactions with each other and with the environment (Epstein and Axtell 1996; Parker et al. 2003; Brown and Duh 2004; Brown et al. 2005). ABMs consist of one or more types of agent embedded in a non-agent environment. Agents may be individuals (e.g., householders, farmers, fishers) or institutions (e.g., a local government, conservation NGOs, firms). The state of an agent can include various characteristics, preferences, and memories of recent events, as well as particular spatial and social connections. Agent definitions include their capabilities to carry out particular behaviors as well as their decision-making rules, heuristics, learning, and other mechanisms, which the agents use to generate their individual behaviors in response to inputs from other agents and from the environment. Often, empirical data are used to establish the initial conditions of the system, specifying the initial attributes of an agent that include type characteristics, intrinsic behavioral rules, modes of communication and learning, and internally stored information about itself and other agents (Tesfatsion 2003). The generation of different types of landscape patterns over space and time based on different theoretical approaches yields a set of future scenarios of change that can include endogenous changes and exogenous shocks and can alter trajectories of landscape change (Rindfuss et al. 2008; Mena et al. 2011).

## A Model of the Galapagos Islands: An Example

Figure 3.1 is a conceptual model of the Galapagos Islands that we have created to demonstrate how complex systems and ABMs can be used to explore human–environment interactions and to test ideas about "what-if" scenarios of social and ecological change. It is a simplified view of the complex population–environment system in the Galapagos Islands. In the picture model, we have indicated several boxes that reflect the primary system components that we will consider. The boxes represent demographic, socioeconomic, and ecological subsystems of the broader system and are explicitly linked through flow arrows that indicate positive (+) or negative (−) relationships. Feedback loops are indicated by arrows that connect two boxes through positive and negative relationships. Between the boxes and associated with an arrow with either a positive (+) or negative (−) effect, we have indicated key processes worthy of empirical analyses to help define rules of behavior and sets of relationships, such as *migration, off-farm employment, agricultural markets, land abandonment,* and *management/genetic repository*. While exogenous factors are important to the behavior of the resource conservation and economic development system in the Galapagos Islands, we have tried to indicate only those that we will consider and then within a general way so that the system can be kept

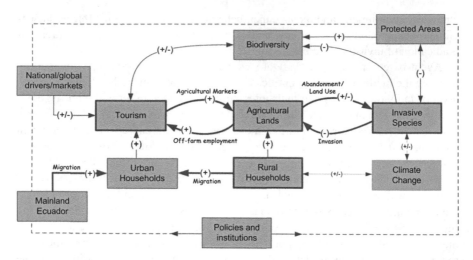

**Fig. 3.1** A generalized system overview of human–environment interactions in the Galapagos Islands of Ecuador

relatively simple to demonstrate the alternative complexity theory context for addressing challenges to the Galapagos Islands. We include international organizations and national (including the Ecuadorian mainland) and local governments in our system as they affect policies, practices, and other institutions (e.g., the United Nations Development Program, Conservation NGOs). We also include climate changes (i.e., El Niño and La Niña events) as they have historically affected populations and biodiversity, the drivers of invasive species, the drivers of global and national markets for marine resources, and the adoption of fisheries as a household livelihood alternative. Further, fisheries and tourism are viewed through the lens of labor (i.e., employment opportunities), considered as part of alternative household livelihood strategies. Of particular interest is the impact of farm abandonment caused by the "push" of invasive plants and the cost and effort of eradication approaches, as well as the general lack of market integration of farmers to sell their products throughout the islands and on the mainland. The "pull" factors include the higher-wage employment opportunities in the tourism and fisheries sectors.

Using an ABM framed within complexity theory, we can spatially simulate population pressures as "shocks" to social–ecological systems to further assess, for instance, the impact of institutions and policies on the behavior of integrated systems. Within our models, we can, for instance, (a) increase the amount of visitors to the islands by 20 and 30 % or more; (b) allow more temporary labor into the islands from the mainland to work on farms and in the construction, tourism, fishing, and service industries; (c) increase the number of tour boat operators and the size of the boats, thereby accommodating more tourists and increasing the vulnerability of the social–ecological system; (d) increase or decrease the presence of tour operators and service providers of transnational companies; (e) vigorously enforce (and relax) fisheries' regulations for local/global operators; (f) reduce the frequency and intensity of El Niño events, thereby reducing the spread of invasive plant species and maintaining household participation in local fisheries that have global implications; and (g) increase the effectiveness of government policies that reduce immigration from the mainland and restrict international tourism.

## A Galapagos Example of Scenario Testing Through an ABM

As an example as to how one can move from the theoretical to the applied, Miller et al. (2010) developed a virtual ABM for the Galapagos Islands—a model that represents key selected elements of the islands to test hypotheses and empirical relationships on system behaviors and dynamics. The ABM was designed to explicitly examine complex and dynamic systems in the study of coupled human–natural systems, with an emphasis on social–ecological interactions in the Galapagos Islands. The virtual environment was designed to maintain the fundamental characteristics of the Galapagos Islands without incorporating needless definition and "noise" in the geographic setting and the modeling environment (Walsh et al. 2009). The ABM examines the challenges of resource conservation and economic development on land use change, including, for instance, the spread of invasive plant species, and alternate household livelihood strategies, particularly employment diversification and job "switching" among the agricultural, fishing, and tourism sectors in the Galapagos, through a set of associated and modeled social–ecological "pushes" and "pulls" that affect human behavior and environmental dynamics. The spatial simulation model was designed to integrate disparate social and ecological data, organized within a GIS, to examine the interactions and feedbacks between people and environment in an island setting. The virtual model most closely resembles Isabela Island, which is geographically positioned in the western portion of the archipelago, populated by approximately 2,500 residents, and is a younger, more volcanically active island in the archipelago. With a coastal community and an agricultural zone in the highlands, residents and tourists move between the two settings to work on family farms and to experience important ecotourism sites, respectively. The virtual Galapagos is similar to actual conditions in that the primary tourism community is located along the coast, agriculture is restricted to the highlands, fisheries are primarily a near-shore activity, and the protected area envelops the human use zones.

Figure 3.2 shows the central elements of the generalized Galapagos model. It is designed to examine land use change—primarily the spread or contraction of the invasive species, guava, in the agricultural highlands—and the ability of island residents to switch between employment sectors in agriculture, fisheries, and tourism relative to new economic opportunities, such as additional jobs in tourism or in response to ecological disincentive, such as a reduction in local fisheries as a consequence of El Niño events. The agents are farmers in the agricultural zone, fishermen in the marine zone, tourism workers in the urban area, park employees in the protected area, and invasive guava that operates on the underlying environmental grid within the agricultural zone and the protected areas. Guava agents are controlled by

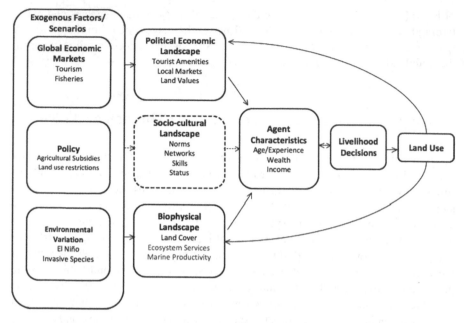

**Fig. 3.2** Model design for a virtual Galapagos Islands (after Miller et al. 2010)

neighborhood effects, meaning that the spread of guava is influenced by the behavior of adjacent farmers and the characteristics of surrounding land parcels. Park agents are only involved in the eradication of guava and are free to move around the entire protected area.

The model acknowledges the importance of exogenous factors in shaping human conditions and the behavior of agents (Gonzalez et al. 2008; Walsh et al. 2011). External factors, such as global market conditions for fisheries and tourism, public policy regarding agricultural subsidies and land use restrictions, and environmental variations including El Niño–Southern Oscillation (ENSO) events and the spread or eradication of invasive species, all have the capacity to impact local socioeconomic (e.g., local market conditions and amenities for tourists), cultural (e.g., social networks and the sharing of information), and biophysical (e.g., marine productivity and ecosystem goods and services) characteristics of the Galapagos Islands. As agents (i.e., individuals and/or households) learn and adapt relative to exogenous forces and endogenous factors, they can choose to diversify their employment patterns by moving between agriculture, fisheries, and tourism, often engaging simultaneously in more than one livelihood alternative in response to dynamic social and ecological conditions. Agent characteristics, such as age, experience, education, wealth, income, and local knowledge combine to influence how social–ecological factors influence their thinking relative to the model outcomes that involve job switching, the accumulation of wealth, and land use change patterns, particularly those linked to the spread or eradication of invasive species on household farms.

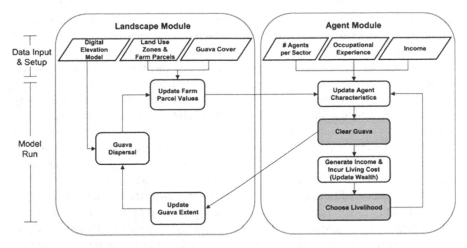

**Fig. 3.3** Landscape Module and Agent Module (after Miller et al. 2010)

The model operates on an annual basis where a graphical user interface allows the analyst to interact with "real" data on farm locations, island population, land use conditions, tourism levels, and the number of licensed fishing boats through a series of levers or switches that increase or decrease variable magnitudes. The intent is to create a social–ecological laboratory in which scenarios can be examined by perturbing a base model. The accumulated wealth of agents is tabulated; job switches among agriculture, fisheries, and tourism are tracked; costs are incurred when agents move from one employment sector to another; trajectories of change in the environment are identified; and interactions between agents are observed as they learn and adapt to changing social and ecological conditions, such as the spread of invasive species on abandoned land, the eradication of invasive species on managed agricultural land, the adoption of public policies to encourage the return of farmers to their household farms through government incentive programs, such as farm subsidies, and the degradation or enhancement of marine and/or terrestrial environments as a consequence of El Niño or La Niña events. In short, the ABM has agents specified in space and time, and the environment is represented on a spatially referenced grid that serves as the physical space of social–ecological interactions of individuals and households with their environment.

Figure 3.3 shows the Landscape Module and the Agent Module (after Miller et al. 2010). The Landscape Module primarily involves the spread of guava. Without data on the spread rate and social–ecological factors that govern the behavior of guava on and off farms in the agricultural highlands, the areal expansion or constriction of guava is calibrated to mirror-observed rates of change from a satellite image time series. A Digital Elevation Model (DEM) is used to characterize terrain settings, important in land use change patterns. In addition, land use zones and farm parcel boundaries are defined through shapefiles organized within a GIS. The Agent Module primarily involves the behavior of farmers, fishermen, tourist industry workers (not tourists per se), and park employees, as well as the household livelihood decisions that the agents make, that is, remaining in or switching to alternative employment opportunities in fisheries, tourism, and agriculture, given financial motivations and ecological opportunities. The initial number of agents is taken from the 2006 population census for the Galapagos. The income of each agent is randomly selected from a livelihood-specific, truncated normal distribution based on the 2006 population census, and gross income is estimated, as is the agent's ability to switch between livelihood options, given potential and actual income levels. Agents can switch livelihoods based on a number of inputs: (1) expected number of tourists, (2) number of open fishing licenses, (3) selling price of fish and agricultural products, (4) start-up costs, and (5) cost of maintaining property. Agents can accumulate wealth according to livelihood decisions. Agent characteristics are updated relative to farm conditions and the areal extent of guava, for example, deciding to stay in agriculture and eradicating guava to enhance crop productivity. The cost of eradicating invasive species is calculated, negatively impacting living costs and household wealth, thereby influencing a farmer's decision to remain in agriculture or switch to fisheries and/or tourism, with associated costs.

Selected results from the hypothetical model indicate the following: the number of people working in fisheries remains stable across the 50 years of the model run, while the number of farmers initially declines as they transition to the tourism sector; the population in tourism and farming remains stable, except for small changes during El Niño years; tourism increases due to switching livelihoods during those years; some farmers "exit" the system due to low accumulated wealth and an inability to switch to alternate sectors; guava initially decreases and then increases, and by the end of the observation period, the area in guava is virtually the same as in the initial conditions. For the scenario involving a government-provided subsidy to promote agriculture in the highlands, the largest farm subsidy ($3,000) has the greatest effect on the population of farmers and the total area in guava; farmers increase in number, and employment patterns remain high throughout the study period; a collapse in global fisheries is accompanied with a decrease in farmers and fishers, while the number of workers in tourism increases and the median income in fisheries and farming decreases; and the area in guava decreases during and after the tourism decline. Lastly, for the scenario that examines a decline in tourism associated with a global economic slowdown, the number of workers in fisheries decreases during the period of low tourism revenues, and the number of farmers increases; as the tourism industry recovers from the economic crisis, the number of tourism workers increases, while the number of farmers decreases to near baseline levels; and guava shows a slight decrease during and after the tourism decline.

## The Tourism System

Presently, tourism and all of its related goods and services form the major economic engine in the Galapagos Islands. The idea of the archipelago as a destination for nature-based tourism that would simultaneously advance economic development and support conservation is difficult to imagine right now. We are currently developing an agent-based model (ABM) to understand the future effects of different trends in tourism as the core feature of the socio-environmental system for the Galapagos Islands. We assume that, in the future, in great part, tourism in the Galapagos will be controlled by global forces, including the types of tourists who are willing to come to the Galapagos Islands and their expectations and willingness to pay for special services.

As we know from both the Galapagos and from other tourist destinations, different types of tourism evolve over time (Vera et al. 1997). Tourism in the Galapagos can be divided into four phases. First, when the tourist sector started to develop in the late 1960s, emphasis was entirely placed on overnight boat tours, a model that was promoted under the pretext of having less of an environmental impact, but in reality, it had more to do with gaining higher revenues for tour operators (Grenier 2007). In addition, vessels with overnight accommodations provided a solution to the lack of infrastructure and services on land (Watkins and Oxford 2008). In the 1980s, a more economical, land-based tourism (i.e., day tours, rather than overnight tours, small hotels, and hostels) began to flourish as tourists from different socio-economic classes started visiting the archipelago; the overnight boat tour model simultaneously continued to develop throughout this decade (Grenier 2007). In the early 1990s, alleged concerns about rapid population growth caused by the expansion of land-based tourism led to an attempt to return exclusively to the overnight boat tour model. What resulted was what Grenier (2007) refers to as "selective tourism" that favors first-class accommodations, often provided by local and overseas tour operators offering overnight tours. With the 1998 enactment of the Special Law, which guaranteed a larger percentage of tourism revenues would remain in the islands; land-based tourism experienced a revival in the early years of the new millennium. More hotels and agencies were created and both land- and sea-based tourism continued to grow (Grenier 2007). In the future, different types of tourism will have different social, economic, and environmental impacts on the Galapagos.

Tourism is a human activity involving multi-scale phenomena that produce multi-scale patterns of impact (Baggio 2008). These impacts are driven at different scales by individual tourists, industries, and communities interacting upon geographical, economic, and ecological conditions. The macroscopic patterns generate feedback processes that, in turn, affect the interaction of tourists, tourism markets, the environment, and local populations. Examples of feedback effects include evolving markets, competition among tourist sites, changes in the attractiveness of tourist sites, and ecological deterioration. Feedback processes play an important role in driving development and bringing about social and ecological changes that produce inherently complex behaviors (Pizzitutti and Mena 2011, unpublished report).

This work in progress is the construction of *GalaSim*, an ABM simulation to capture decision-making at the very basic unit, that is, the individual tourist at different stages of the travel experience. The tourist chooses from a portfolio of opportunities, in some cases, long before the trip actually starts. These decisions are based upon the characteristics of the trip and the destination but are also based on the interests and socioeconomic characteristics of the tourist. Tourists are parameterized

**Fig. 3.4** Diagram of the *GalaSim* model (Pizzitutti and Mena 2011, unpublished report)

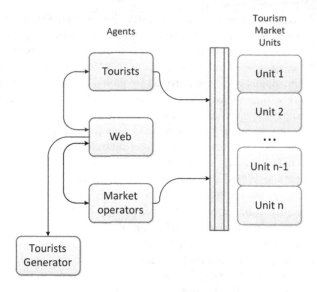

using characteristics collected by the official registry system of the Galapagos National Park. The model also aims to capture current local tourism infrastructure and markets and, in the future, implications to the local economy and environment. It is important to note that the use of an ABM to model tourism has been very limited (Cecchini and Trunfio 2007; Johnson and Sieber 2011), mainly due to the lack of information about tourism systems. In Galapagos, the rich collection of data about tourism and tourism infrastructure is important for the development of this kind of model.

In the first part of this tourism study, the model captures the interactions between tourists and tourism market operators in the Galapagos. In this model, market operators and tourists interact through a "virtual" platform, such as the Internet, that represents the interaction space for negotiating (a) the sale promotion, (b) the sharing of information about products, and (c) the purchase of many of the tourists' selections. This virtual platform can be viewed as a place for the storage of structured knowledge that permits tourists and market operators to engage each other and for the tourist to make trip selections and even trip preparations. The tourism market is represented as units that correspond to the tourism offerings and that can be viewed as a cell where tourists are allocated (Pizzitutti and Mena 2011, unpublished report).

Tourists enter into the system at rates corresponding to the official statistics of the Galapagos National Park and are created by a tourist generator agent, following a psychographic distribution (Plog 2001), and other distribution curves that describe demography, awareness, preferences, and budget (Pizzitutti and Mena 2011, unpublished report) (Fig. 3.4). The model uses a matrix of priorities based on endogenous characteristics and preferences. Tourists then choose market units, distributed across Santa Cruz, San Cristobal, and Isabela islands. Important agents are market operators who have the capacity to organize tourist agents. Market operators differ in terms of knowledge, access to information, efficiency, quality of products, and money to invest in tourism and their services. Tourist agents who choose vessel- and land-based modes are treated separately. The aim is to understand future trends of mobility within the archipelago and, eventually, the ecological and economic implications of tourism in the islands on human–environment interactions and the generation of income, wealth, and assets.

## Climate Change as an Agent of Change

In the above ABM models, climate change is examined indirectly by creating linked relationships between the spread of invasive species and the choice of household livelihoods in agriculture, fisheries, and tourism. Climate models for the Eastern Pacific generally indicate that El Niño events will likely increase in frequency and magnitude for the Galapagos Islands. As such, ABMs can be used to model their impacts as social–ecological shocks to the social, terrestrial, and marine subsystems of the Galapagos.

In the Galapagos, the opposite extremes of ENSO, El Niño, and La Niña events have strong and contrasting implications for the stability of native ecosystems (Bliemsrieder 1998), the spread of invasive species (Tye and Aldaz 1999), and human livelihoods (Cruz 1985; Robalino et al. 1985). El Niño and La Niña events and their effects on terrestrial ecosystems are relatively well studied (e.g., Holmgren et al. 2001), although many uncertainties remain, mostly related to spatial and temporal lags. In the Galapagos Islands, reports of the effects of El Niño date back to the early 1950s, but the events that occurred in 1982–1983 and 1997–1998 have been well described, dealing mostly with the effects on endemic populations of flora and fauna.

The increase in rainfall associated with El Niño in Western South America has been identified as an important trigger for seed germination and germination blooms (Arntz and Fahrbach 1996). There are indications, however, that herbs are more sensitive than shrubs (Jaksic 2001) and that the effects differ over small spatial scales that increase patchiness in primary production (Gutiérrez and Meserve 2003; Jaksic 2001). In the Galapagos, reports indicate that the elevated rainfall rates and totals increase woody tree mortality, especially in dry areas; support the spread of selected invasive species; and promote the expansion of existing lianas and grasslands (Hamann 1985; Luong and Toro 1985; Itow 2003). Increased rainfall increases primary productivity by first promoting the germination of dormant seeds that later are consumed and spread by herbivores. Reports indicate that El Niño contributes to the appearance and spread of invasive species (Tye and Aldaz 1999). The link between herbivores and the spread of guava also appears to be quite strong, as seeds are dispersed by cattle, horses, pigs, birds, and rats (Ellshoff et al. 1995; GISD 2005) that eat the fruits and excrete the numerous seeds.

El Niño also affects the livelihoods of humans in the Galapagos. Reports point out the important decrease in fishing stocks (Robalino et al. 1985) and the negative effect on agricultural activities, including flooding pastures, destruction of cash and subsistence crops, diseases in cattle, and impacts to the infrastructure (Cruz 1985; Robalino et al. 1985). Fishermen and farmers must adapt to the changing environment, and in the case of El Niño, they must adapt to cope with this exogenous shock as there are strong feedbacks between land use intensity, land abandonment, and the effects of El Niño.

Conflicts between resource conservation and economic development in the Galapagos Islands occur as a consequence of a burgeoning human migrant population, primarily from the mainland of Ecuador and from tourists who visit the archipelago from around the world. This growing human population is now threatening the future of this ecologically fragile area. Due to this, in April 2007, the United Nations designated the Galapagos Islands "at risk" from the threats associated with population growth and economic development. Similarly, the Ecuadorian government declared an "ecological emergency" in the world-renowned Galapagos National Park and Marine Reserve.

One of the greatest threats to the ecosystems of the Galapagos Islands is the growing number, severity, and areal expansion of exotic plant and animal species (Tye et al. 2002; Tye 2006). Increased human presence has hastened the introduction of invasive species that are now so prevalent and severe that they threaten the native and endemic flora and fauna of the islands, ecosystem services, and the human–natural system. Thirty-seven of more than 800 alien plant species are considered highly invasive in the Galapagos Archipelago (Tye et al. 2002). Relative to the number of species they endanger, exotic species are the least studied threat to biodiversity (Lawler et al. 2006).

## Measuring Landscape Dynamics

While ABMs are capable of spatially simulating shifts in human behavior and the adaptive resilience of social and ecological systems to environmental change, it is important to characterize initial conditions for the onset of the model and to correctly represent the composition and spatial pattern of land use/land cover for the study area, as well as the social and ecological landscapes that are fused together through an *Island Biocomplexity* context.

While information exists on all tourists and temporary workers who enter the Galapagos, a separate system tracks the entries and exits of all permanent and temporary residents. Ecuadorian census data were collected for 1990, 1998, 2001, 2006, and 2010, and a Living Standards survey was conducted for the islands in 2009 that provides detailed information on a wide variety of economic activities. These data are critical for establishing many of the rules and relationships used in ABMs, particularly, the geo-location of dwelling units, census units, demographic characteristics, roads, land parcels, and associated information that is used to characterize social dimensions in the islands. Conversely, satellite systems have been increasingly relied upon to gather space-time information on the environment, particularly land use/land cover change, as well as sea surface temperatures and the chlorophyll content of the marine environment. We have implemented strategies that involve the fused use of high spatial resolution data (e.g., WorldView-2, QuickBird, Ikonos, or ADAR digital aircraft data) fused with moderate resolution imagery (e.g., ASTER, Landsat), as well as coarse-grained systems such as MODIS imagery. In addition, we have fused optical systems with non-optical radar systems for landscape characterization.

In addition, analyses have been conducted that use multispectral satellite data such as Landsat (e.g., Joshi et al. 2006; Huang and Zhang 2007) and Advanced Land Imager data (ALI) (e.g., Stitt et al. 2006) versus the use of hyper-spectral data such as Hyperion (e.g., Asner et al. 2006; Pengra et al. 2007; Underwood et al. 2007; Walsh et al. 2008), and hyper-spectral digital aircraft data (e.g., Underwood et al. 2003; Miao et al. 2006; Hunt and Parker-Williams 2006) for characterizing land use/land cover change patterns. For Isabela, for instance, aerial photography was collected in 1959/1960, 1982–1985, 1992, and 2007. The imagery is maintained by the Ecuadorian Geographic Military Institute (IGM) for all of the Galapagos Islands. The March 2007 mission characterized the landscape of the Galapagos Islands at a scale of 1:30,000, in natural color, and with standard forward- and side-lap for stereoscopic viewing.

The general design is to acquire historical aerial photography and spatial-, temporal-, and spectral-resolution satellite data to construct a trend analysis of land use/land cover change and plant invasions. It is common to fuse multiple data sets, such as hyper-spectral Hyperion and multispectral Advanced Land Imager data. Historical to contemporary satellite imagery—including Landsat Thematic Mapper, Landsat Multispectral Scanner, and ASTER—as well as the 2007 natural color aerial photography of the Galapagos Islands and earlier aerial photo mission data are used as well. The temporal coherence of the imagery across the various sensor systems and image dates can be maintained.

A processing template can be developed that includes a consistently applied set of image preprocessing operations to spectrally, geometrically, and radiometrically correct images from each sensor system and time series. Preliminary steps often include the generation of a consistent set of vegetation indices (e.g., Normalized Difference Vegetation Index, Soil Adjusted Vegetation Index, Fractional Cover Index) and the Tasseled Cap Wetness-Greenness-Brightness transforms to extend the feature sets for image classification. Primary analyses can be based on pixel-based approaches (e.g., unsupervised, supervised) and object-based image analyses (OBIA) approaches to characterize the landscape into general land use/land cover types, with special emphasis on mapping forest (degraded and otherwise), grasslands, cropland, pasture, bare soil, and invasive species. Walsh et al. (2008) examined the use of multispectral QuickBird data and hyper-spectral Hyperion data to characterize guava for a test area on Isabela Island. Findings indicate a positive synergism between the different types of data and different image-processing methods (i.e., linear vs. nonlinear spectral unmixing and pixel vs. object-based image analysis) to characterize the environment.

# Conclusions

Operating within an *Island Biocomplexity* context, we advocate a framework and perspective that is capable of addressing the linked effects of social–ecological systems in the Galapagos Islands. *Island Biocomplexity* maintains an adaptive resilience of local factors and distal forces that function through the coevolution of human–environment interactions to understand complex island ecosystems (Michener et al. 2002). Further, we demonstrate how an agent-based model can be developed to examine various scenarios of change to the social, terrestrial, and marine subsystems of the Galapagos Islands by fusing social and ecological information from social surveys and a satellite time series to develop rules and relationships, and a rich process understanding, of complex and dynamic systems in the Galapagos Islands and beyond.

The use of *Island Biocomplexity*, or complexity theory, within island settings offers a great potential for understanding coupled human–environmental systems, mainly through the generation of input and output parameters, such as flows of people, material, and capital. This is true in the Galapagos Islands, where relatively good information exists to describe the social and environmental domains. Additionally, ABMs and other methodological tools based on complexity explicitly embrace uncertainty as part of the system that is key for environmental management in island ecosystems, where small variations in key variables can change the trajectories and conditions of entire social–ecological systems.

Here, we have described work we are conducting using the Galapagos Islands as a natural laboratory. These examples illustrate a range of ABM applications from agriculture to tourism that can create future scenarios relevant to policy. Although complex systems research, including ABMs, is expanding quickly in the social and natural sciences, the methods are still experimental, and applications to public policy making are relatively few in number. The Galapagos Paradox can be tested using complex systems, but models can only inform about possible future scenarios and operative pattern–process relations. It is human agency that must protect this very charismatic and amazing place. *Island Biocomplexity* and complex adaptive systems are new frameworks and perspectives to assess the challenges of the Galapagos Islands and to present plausible alternative futures to protect and preserve this magical place.

# References

Arntz W, Fahrbach E (1996) El Niño: experimento climático de la naturaleza. Fondo de Cultura Económica, Mexico

Asner GP, Martin RE, Carlson KM, Rascher U, Vitousek PM (2006) Vegetation–climate interactions among native and invasive species in Hawaiian rainforest. Ecosystems 9:1106–1117

Baggio R (2008) Symptoms of complexity in a tourism system. Tourism Anal 13(1):1–20

Bliemsrieder M (1998) El Fenomeno de El Niño en Galapagos. Informe Galápagos 1997–1998. Fundacion Natura–World Wildlife Fund, Quito, Ecuador, pp 48–50

Brown DG, Duh JD (2004) Spatial simulation for translating between land use and land cover. Int J Geogr Inf Sci 18(1):35–60

Brown DG, Riolo R, Robinson DT, North M, Rand W (2005) Spatial process and data models: toward integration of agent-based models and GIS. J Geogr Syst 7(1):25–47

Cecchini A, Trunfio GA (2007) A multi-agent model for supporting tourism policy-making by market simulations. Int Conf Comput Sc 1:567–574

Cilliers P (1998) Complexity and postmodernism. Routledge, New York

Cruz E (1985) Efectos del Niño en la Isla Floreana. In: Robinson G, Del Pino E (eds) El Niño en las Islas Galápagos: Evento. Charles Darwin Foundation, Galapagos, Ecuador, pp 1982–1993

Ellshoff ZE, Gardner DE, Wikler C, Smith CW (1995) Annotated bibliography of the genus *Psidium*, with emphasis on *P. Cattleianum* (strawberry guava) and *P. Guajava* (common guava), forest weeds in Hawaii. Technical report 95, Cooperative National Park Resources Study Unit, University of Hawaii at Manoa

Epstein JM, Axtell R (1996) Growing artificial societies: social science from the bottom up. MIT Press, Cambridge, MA

Evans TP, Kelley H (2004) Multi-scale analysis of a household level agent-based model of land-cover change. Environ Manage 72(1–2):57–72

Global Invasive Species Database (GISD) (2005) *Psidium guajava.* http://www.issg.org/database/species/ecology.asp?si=211&fr=1&sts=. Accessed June 2, 2012. Last modified 11 Apr 2006

Gonzalez JA, Montes C, Rodríguez J, Tapia W (2008) Rethinking the Galapagos Islands as a complex social–ecological system: implications for conservation and management. Ecol Soc 13(2):13 (online)

Grenier C (2007) Galápagos, conservación contra natura. Quito: Abya-Yala, Universidad Andina Simón Bolivar, IFEA, IRD, Coopération Française, pp 463

Gutiérrez JR, Meserve PL (2003) El Niño effects on soil seed bank dynamics in north-central Chile. Oecologia 134(4):511–517

Hamann O (1985) The El Niño influence on the Galapagos vegetation. In: Robinson G, Del Pino E (eds) El Niño en las Islas Galápagos: Evento. Charles Darwin Foundation, Galapagos, Ecuador, pp 1982–1993

Holmgren M, Scheffer M, Ezcurra E, Gutierrez JR, Mohren GMJ (2001) El Niño effects on the dynamics of terrestrial ecosystems. Trends Ecol Evol 16(2):89–94

Huang H, Zhang L (2007) A study of the population dynamics of *Spartina alterniflora* at Jiuduanmsha Shoals, Shanghai. Chin Ecol Eng 29:164–172

Hunt ER Jr, Parker-Williams AE (2006) Detection of flowering leafy spurge with satellite multi-spectral imagery. Rangel Ecol Manage 59(5):494–499

Itow S (2003) Zonation pattern, succession process, and invasion by aliens in species-poor insular vegetation of the Galapagos Islands. Global Environ Res 7(1):39–58

Jaksic FM (2001) Ecological effects of El Niño in terrestrial ecosystems of western South America. Ecography 24(3):241–250

Johnson PA, Sieber RE (2011) An agent-based approach to providing tourism planning support. Environ Plann B 38:486–504

Joshi C, De Leeuw J, van Andel J, Skidmore AK, Lekhak HD, van Duren IC, Norbu N (2006) Indirect remote sensing of a cryptic forest understory invasive species. For Ecol Manage 225:245–256

Lawler JL, Aukema JE, Grant JB, Halpern BS, Kareiva P, Nelson CR, Ohleth K, Olden JD, Schlaepfer MA, Silliman BR, Zaradic P (2006) Conservation science: 20-year report card. Front Ecol Environ 4(9):473–480

Luong TT, Toro B (1985) Cambios en la vegetacion de las Islas Galapagos durante "El Niño" 1982–1983. In: Robinson G, Del Pino E (eds) El Niño en las Islas Galápagos: Evento. Charles Darwin Foundation, Galapagos, Ecuador, pp 1982–1993

Malanson GP (1999) Considering complexity. Ann Assoc Am Geogr 89(4):746–753

Malanson GP, Zeng Y, Walsh SJ (2006) Complexity at advancing ecotones and frontiers. Environ Plann A 38:619–632

Manson SM (2001) Simplifying complexity: a review of complexity theory. GeoForum 32(3):405–414

Matthews KB, Subaald AR, Craw S (1999) Implementation of a spatial decision support system for rural land use planning: integrating geographic information systems and environmental models with search and optimization algorithms. Comput Electron Agric 23:9–26

Mauchamp A (1997) Threats from alien plant species in the Galapagos Islands. Conserv Biol 11(1):260–263

Mena CF, Walsh SJ, Frizzelle BG, Malanson GP (2011) Land use change of household farms in the Ecuadorian Amazon: design and implementation of an agent based model. Appl Geogr 31(1):210–222

Miao X, Gong P, Swope SM, Pu R, Carruthers RI, Anderson GL (2006) Estimation of yellow starthistle abundance through CASI-2 hyperspectral imagery using linear spectral mixture models. Remote Sens Environ 101(3):329–341

Michener WK, Baerwald TJ, Firth P, Palmer MA, Rosenberger JL, Sandlin EA, Zimmerman H (2002) Defining and unraveling biocomplexity. Bioscience 51(12):1018–1023

Miller BW, Breckheimer I, McCleary AL, Guzman-Ramirez L, Caplow SC, Walsh SJ (2010) Using stylized agent-based models for population–environment research: a case from the Galapagos Islands. Popul Environ 31(6):401–426

Parker DS, Manson SM, Janssen M, Hoffmann M, Deadman P (2003) Multi-agent systems for the simulation of land use and land cover change: a review. Ann Assoc Am Geogr 93(2):314–337

Pengra BW, Johnston CA, Loveland TR (2007) Mapping an invasive plant, *Phragmites australis*, in coastal wetlands using the EO-1 Hyperion hyperspectral sensor. Remote Sens Environ 108:74–81

Pizzitutti F, Mena CF (2011) GalaSim: an Agent Based Model simulation for touristic flows in the Galapagos Islands. Unpublished report. Universidad San Francisco de Quito, Ecuador

Plog SC (2001) Why destination areas rise and fall in popularity? Cornell Hotel Rest Q 42(3):13

Rindfuss RR, Entwisle B, Walsh SJ, An L, Badenoch N, Brown DG, Deadman P, Evans TP, Fox J, Geoghegan J, Gutmann M, Kelly M, Linderman M, Liu J, Malanson GP, Mena CF, Messina JP, Parker DC, Robinson D, Sawangdee Y, Verburg P, Zhong G (2008) Land use change: complexity and comparisons. J Land Use Sci 3(1):1–10

Robalino E, Goumaz L, Alvear L, Ciza A, Castañeda L, Martinez W, Patiño M, Lopez M (1985) Effectos del Niño 1982–1983 en el Hombre y sus actividades en la Isla Santa Cruz, Galápagos. In: Robinson G, Del Pino E (eds) El Niño en las Islas Galápagos: Evento. Charles Darwin Foundation, Galapagos, Ecuador, pp 1982–1993

Stitt S, Root R, Brown K, Hager S, Mladinich C, Anderson GL, Dudek K, Bustos MR, Kokaly R (2006) Classification of leafy spurge with Earth observing-1 advanced land imager. Rangel Ecol Manage 59:507–511

Tesfatsion L (2003) Agent-based computational economics: modeling economies as complex adaptive systems. Inform Sci 149(4):262–268

Tye A (2006) Can we infer island introduction and naturalization rates from inventory data? Evidence from introduced plants in Galapagos. Biol Invasions 8:201–215

Tye A, Aldaz I (1999) Effects of the 1997–98 El Niño event on the vegetation of Galápagos. Not Galápagos 60:25–28

Tye A, Soria MC, Gardener MR (2002) A strategy for Galápagos weeds. In: Veitch CR, Clout MN (eds) Turning the tide: the eradication of invasive species. IUCN SSC Invasive Species Specialist Group, IUCN, Gland, Switzerland and Cambridge, UK

Underwood E, Ustin SL, DiPierto D (2003) Mapping non-native plants using hyperspectral imagery. Remote Sens Environ 86(2):150–161

Underwood EC, Ustin SL, Ramirez CM (2007) A comparison of spatial and spectral image resolution for mapping invasive plants in coastal California. Environ Manage 39:63–83

Vera JF, López F, Marchena M, Anton S (1997) Análisis territorial del turismo. Arial Geografía, Barcelona

Walsh SJ, McCleary AL, Mena CF, Shao Y, Tuttle JP, Gonzalez A, Atkinson R (2008) QuickBird and Hyperion data analysis of an invasive plant species in the Galapagos Islands of Ecuador: implications for control and land use management. Remote Sens Environ Spec Issue Earth Observ Biodivers Ecol 112:1927–1941

Walsh SJ, Mena CF, Frizzelle BG, DeHart JL (2009) Stylized environments and ABMs: educational tools for examining the causes and consequences of land use/land cover change. GeoCarto Int 24(6):423–435

Walsh SJ, Malanson GP, Messina JP, Brown DG, Mena CF (2011) Biocomplexity. In: Millington A, Blumler M, Schickhoff U (eds) The SAGE handbook of biogeography. SAGE Publications, London, pp 469–487

Watkins G, Oxford P (2008) Galapagos: both sides of the coin. Enfoque Ediciones, Quito

# Chapter 4
# The Socioeconomic Paradox of Galapagos

Byron Villacis and Daniela Carrillo

## Introduction

The Galapagos Islands are an Ecuadorian province located in the Eastern Pacific Ocean about 1,000 km off the mainland of Ecuador. The Galapagos archipelago is composed of 13 large islands, 6 small islands, and 107 rocks and islets. These islands were made famous by Charles Darwin's theory of evolution by natural selection[1] and by the presence of numerous endemic species. The Galapagos Islands are called the "Enchanted Islands," because of their unique flora and fauna, which are almost impossible to replicate in other regions around the globe.

The Galapagos Islands are an inspiration for research in social and ecological sciences (Tapia et al. 2009).[2] There has been very little social science research conducted in the islands compared to the enormous amount of natural sciences research. This paucity of research on the human dimension in the Galapagos Islands has contributed to a general lack of understanding about the links between natural and human ecosystems. The goal of this chapter is to provide supporting evidence for the increasing importance of social processes in shaping the Galapagos Islands and altering the

---

[1] In 1859, Charles Darwin published his theory of evolution by natural selection as an explanation for adaptation and speciation. He defined natural selection as the "principle by which each slight variation [of a trait], if useful, is preserved." The concept was simple but powerful: individuals that are best adapted to their environments are more likely to survive and reproduce. As long as there is some variation between individuals, there will be an inevitable selection with the most advantageous variations. If the variations are inherited, then differential reproductive success will lead to a progressive evolution of particular populations of a species, and populations that evolve to be sufficiently different eventually will become different species.

[2] Tapia W, Ospina P, Quiroga D, Gonzalez JA, Montes C (2009) Science for Galapagos: a proposed strategy and priority research agenda for sustainability of the archipelago. Quito, Ecuador. http://www.galapagospark.org/documentos/Ciencia_para_la_sostenibilidad_Tapia_et_al_2009.pdf

B. Villacis (✉) • D. Carrillo
National Institute of Statistics and Census Institute – INEC, Quito, Ecuador
e-mail: byronvillacis@gmail.com; danicarrilloc@hotmail.com

S.J. Walsh and C.F. Mena (eds.), *Science and Conservation in the Galapagos Islands:* 69
*Frameworks & Perspectives*, Social and Ecological Interactions in the Galapagos Islands 1,
DOI 10.1007/978-1-4614-5794-7_4, © Springer Science+Business Media, LLC 2013

integrity of natural ecosystems, thereby challenging the conservation paradigms in the "Enchanted Islands." Empirically, we present transversal statistics to expose the paradox of sustainability in Galapagos. Sustainability is understood as an integrated notion between human needs, satisfying their needs without compromising the needs of future generations (World Commission on Environment and Development 1987). We label this situation as a paradox, due to the contraposition between the exits of the present and future problems and the needs of its present population. The principal aim is to prove the paradox of Galapagos: a healthy place to live where the only way to preserve it is to cut the present health. The arguments for an explicit human–environment discourse in the Galapagos Islands are divided into four sections that contain statistical data and demographic analysis: (1) demographic, (2) socioeconomic, (3) health, and (4) conclusions.

## Demographic Analysis

The population growth in Galapagos is becoming untenable. According to the last population census in 2010, there were 25,124 inhabitants of the Galapagos Islands. The intercensal growth rate is 3.32%, with a density of 80 persons per $km^2$. In Ecuador, the growth rate is 1.95% with a density of 56 persons per $km^2$. Figure 4.1 shows the population evolution on the islands compared with the entire country.

Galapagos has a territory of 8,010 $km^2$, where 3.3% is available for human activity and the remaining 96.7% is under the jurisdiction of the Galapagos National Park (2011) and is reserved for the natural ecosystems of the islands. The residential population of Galapagos is principally located on three islands: Santa Cruz, San Cristobal, and Isabela. Table 4.1 shows the population structure, according to municipalities and parroquias (parishes).

Population immigration is the central factor that describes the demography of the archipelago. According to the 2010 census, nearly 60% of the residential population in the Galapagos was born outside of the province, a trend observed over the previous 20 years (Table 4.2).

The immigration problem is related to the informality of the current process of accepting new workers into the islands. Official reports show that local government is increasingly granting residency status. The only way to formalize legal permanence in the Galapagos is to obtain residency, which allows a person to work, study, and use all local services. Before the *Special Law for Galapagos* was issued in 1998,[3] local governments did not have a proper registration system for permanent residents, there was no official process to obtain residence cards and several political and administrative problems existed. It is likely that fraudulent mechanisms were used to facilitate granting permanent residency to people who did not meet legal requirements. The problem still exists as, while the local government has improved the planning, control, and registration of actual and future residents, there

---

[3] This law seeks to regulate the Special Regime for Galapagos and to regulate the legal and administrative elements http://www.ambiente.gob.ec/proyectos/userfiles/51/file/turismo/ley%20galapagos.pdf.

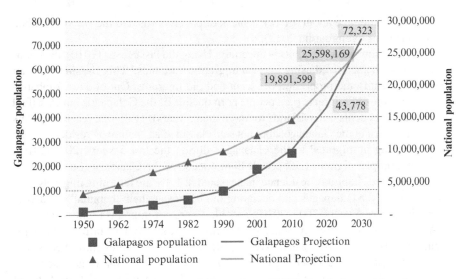

**Fig. 4.1** National and Galapagos population projects (INEC, Population and housing census, 2010a)

**Table 4.1** Population distribution in the Galapagos Islands (INEC, Population and housing census, 2010a)

| Municipality | Parroquias | Population | % Distribution |
|---|---|---|---|
| San Cristobal | Puerto Baquerizo Moreno | 6,672 | 26.6 |
| | El Progreso | 658 | 2.6 |
| | Isla Santa Maria (Floreana) | 145 | 0.6 |
| | Total | 7,475 | 29.8 |
| Isabela | Puerto Villamil | 2,092 | 8.3 |
| | Tomas De Berlanga | 164 | 0.7 |
| | Total | 2,256 | 9.0 |
| Santa Cruz | Puerto Ayora | 11,974 | 47.7 |
| | Bellavista | 2,425 | 9.7 |
| | Santa Rosa | 994 | 4.0 |
| | Total | 15,393 | 61.3 |
| Galapagos | Total | 25,124 | 100.0 |

**Table 4.2** Structure of the population in the Galapagos Islands (INEC, Population and housing census, 2010a)

| Census | Born in Galapagos (%) | Born outside (%) |
|---|---|---|
| 1990 | 35.7 | 55.4 |
| 2001 | 34.5 | 60.6 |
| 2010 | 34.6 | 59.5 |

is a lack of confidence about the correctness of past records and the way to validate previous actions (Vanguardia 2012).[4]

The actual number of migrants is uncertain. Using the last census (2010), the population can be characterized using internal migration variables.[5] This classification can be used to identify the proportion of recent and old migrants. Out of 21,077 inhabitants,[6] 49.3% are old migrants (i.e., people born outside of the Galapagos but classified as "habitual" residents of the islands for more than 5 years); 11.3% are recent migrants (i.e., people born in the Galapagos, but not self-classified as "habitual" residents of the islands, having immigrated to the Galapagos during the last 5 years, 2005–2010); 3.7% are multiple migrants (i.e., people born outside of the Galapagos but who have not self-declared "habitual" residence in the islands); 1.4% are returning migrants (i.e., people born in the Galapagos and declaring the islands as their "habitual" residence, but not in the last 5 years, 2005–2010); and 34.2% are not migrants (Table 4.3).

Additional quantitative evidence of the immigration problem in the Galapagos Islands comes from marriage and divorce statistics. Both variables show a considerable incremental increase between the years 2007 and 2010: marriages increased 511%, from 38 to 232 marriages per year, and divorces increased from 0 to 64 divorces per year (Fig. 4.2). Migrants may be getting married illegally in order to obtain legal status in the islands.[7]

The internal population is not growing at the same rate as immigration. In the Galapagos, births decreased between 2001 and 2010, as the birth rate decreased from 22.7 births per 1,000 inhabitants to 14.1 per 1,000 inhabitants, while the mortality rate slightly increased: 1.6 deaths per 1,000 inhabitants in 2001 compared to 1.8 deaths per 1,000 inhabitants in 2010.[8] Additionally, according to the 2009 Life Condition Survey (2009a),[9] the proportion of pregnant women in the Galapagos was lower than in the whole of Ecuador: 5.7% compared to 6.9%.

Tourism is the apparent driving force behind population growth in the Galapagos Islands, likely pulling new migrants to the islands as well. As the number of foreign and national tourists increases, tourism has become the islands' main economic activity. The flow of tourists in the past decade has drastically increased. Between 2001 and 2011, the total number of tourists increased 138.5%, from 77,570 to 185,028 and—if the number of tourists who came to Galapagos in 1979 is compared—the increase is from 11,765 to 185,028, which represents an increase of 1,472.7% (Fig. 4.3). The correlation between tourists and the population is 0.97.

---

[4] Vanguardia (2012) The Galapagos Report

[5] It is based on the methodology proposed by CELADE to characterize internal migration. This procedure is an international standard for characterizing internal movements.

[6] For methodology, we remove foreigners from the people registered in the census, resulting in 21,077 habitants.

[7] According to Article 26, paragraph 2 of the Special Law for Galapagos Province, the following are permanent residents: "The Ecuadorians or foreigners, who have legalized their stay in the country, maintain spousal or de facto union recognized under the Act or the children of a permanent resident in the province of Galapagos."

[8] National Institute of Statistics and Censuses (2010) Yearbook of vital statistics

[9] The survey was implemented on mainland Ecuador in 2005–2006 and in the Galapagos Islands in 2009.

**Table 4.3** Characterization of immigrants to the Galapagos Islands [Latin American and Caribbean Demographic Center (CELADE), 2010]

|  | Old migrants | Recent migrants | Multiple migrants | Return migrants | Not migrants |
|---|---|---|---|---|---|
| Born in Galapagos | No | Yes | No | Yes | Yes |
| Habitual resident of Galapagos | Yes | No | No | Yes | Yes |
| Habitual resident of Galapagos in the last 5 years, 2005–2010 | Yes | Yes | No | No | Yes |

**Fig. 4.2** Evolution of marriages and divorces in the Galapagos Islands (INEC, Yearbook of vital statistics, 2010d)

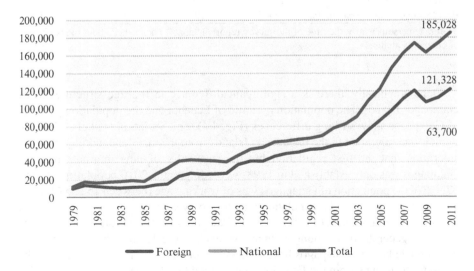

**Fig. 4.3** Historical number of visitors who entered the protected areas of the Galapagos Islands, 1979 through 2011 (Galapagos National Park, 2012)

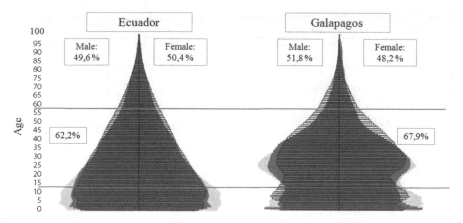

**Fig. 4.4** Population pyramids for Ecuador and the Galapagos Islands showing differences in their sex and age structures (INEC, Population and housing census, 2010a)

**Table 4.4** Population characteristics by place of birth (INEC, Population and housing census, 2010a)

| | Galapagos | | | |
|---|---|---|---|---|
| Variables | Born in Galapagos | Born in other provinces | Foreign-born | Ecuador |
| Average age | 22 | 28 | 31 | 28 |
| Men (%) | 50.7% | 49.6% | 51.0% | 49.6% |
| People of Working age (%) | 48.5% | 50.7% | 59.5% | 43.1% |
| Average household size | 3.7 | 3.8 | 3.4 | 3.8 |
| Single people (%) | 45.8% | 36.5% | 30.5% | 36.5% |
| People without any education[a] (%) | 0.8% | 5.0% | 3.8% | 5.0% |

[a]Refers to persons who at the time of the survey reported "none" when asked about the peak level of education attained

Demographically, there is a significant difference in the age structure in the Galapagos Islands. The population pyramid of Ecuador and the population pyramid of Galapagos show striking differences (Fig. 4.4). This variation is likely related to Ecuadorian migration to the Galapagos Islands from the mainland, taking into consideration that the first regulation limiting the entry of new migrants into the Galapagos was only established in 1998. According to the 2010 population census, 67.9% of the population of Galapagos was of working age (between 15 and 65 years old), and only 4.4% of the population was over 65 years old.

The population currently living in the Galapagos that was born outside of the islands is generally quite young, with an average age greater than the native population. People born in Galapagos have better educational conditions than others who come to the islands from the Ecuadorian mainland (Table 4.4).

In this first section, we have reported statistics on the demographic characteristics of the population in the Galapagos Islands. It is important to highlight that the growth of the total population in the Galapagos was motivated by migration as a principal component and its relationship to tourist activities. In the next section, we will describe socioeconomic conditions in the Galapagos Islands.

## Socioeconomic Analysis

New migrants are motivated by positive socioeconomic conditions in the Galapagos. Job opportunities, particularly in tourism, create the need for services, which, in turn, encourages more displacements and attracts more and more people to the islands (World Wildlife 2003).[10] Compared to the mainland of Ecuador, Galapagos has attractive employment conditions, such as higher incomes, greater access to technology and higher education, and greater gender equity as "pull" factors. To migrate to the Galapagos, people have to confront the *Galapagos Paradox*. The Galapagos has "push" factors as well—a poor educational infrastructure, lack of basic services, violence against women, and relatively high prices for basic food and services. Here, we describe favorable and unfavorable socioeconomic conditions in the islands as "pull" and "push" factors of population migration to the islands.

### *Favorable Socioeconomic Conditions in the Galapagos*

If you live in the Galapagos, it is much easier to find a job, than if you live on the mainland of Ecuador. In 2009, the unemployment rate for Galapagos was 4.9%, while on the mainland of Ecuador the rate was 7.9%. The subemployment was 38.7%, while on the mainland of Ecuador the rate was 50.5%. The fully employed rate in the Galapagos was 64.7%, while on the mainland of Ecuador the rate was 38.8%. An additional consideration is the size of the labor market. According to the 2010 population census, the "economically active population" in the Galapagos was 12,975 persons (i.e., 51.6% of the total population), while on the mainland of Ecuador the "economically active population" was 6,093,173 persons (i.e., 42.1% of the total population). Additionally, the labor market in the Galapagos, as compared to the mainland of Ecuador, has a higher economic participation (i.e., 70.3% versus 67.7%).

In addition, a job in the Galapagos, generally, has higher wages. For instance, any person employed in the public sector in the Galapagos receives double the

---

[10] World Wildlife Fund (2003) Migration and environment in the Galapagos Islands. Quito, Ecuador

**Table 4.5** Market labor structure by economic activity (INEC, Population and housing census, 2010a)

| Economic activity | Ecuador (%) | Galapagos (%) |
|---|---|---|
| Wholesale and retail | 18.4 | 12.8 |
| Public administration and defense | 4.1 | 10.7 |
| Hosting activities and food service | 3.8 | 9.5 |
| Agriculture, livestock, forestry, and fishing | 21.8 | 9.0 |
| Construction | 6.5 | 7.5 |
| Transport and storage | 5.2 | 7.0 |
| Administrative and support services | 2.7 | 7.0 |
| Teaching | 5.1 | 5.6 |
| Manufacturing industries | 10.2 | 5.1 |
| Other activities | 22.1 | 25.8 |
| Total | 100.0 | 100.0 |

**Table 4.6** Economic activity by sector (INEC, National economic census, 2010b)

| Economic activity | Mainland Ecuador | | Galapagos | |
|---|---|---|---|---|
| | Business | % | Business | % |
| Wholesale and retail trade | 269,751 | 53.9 | 545 | 41.6 |
| Hosting and food service | 51,815 | 10.4 | 247 | 18.9 |
| Other service activities | 39,631 | 7.9 | 105 | 8.0 |
| Manufacturing | 47,867 | 9.6 | 89 | 6.8 |
| Public administration and defense | 4,009 | 0.8 | 54 | 4.1 |
| Transport and storage | 5,228 | 1.0 | 51 | 3.9 |
| Information and communication | 19,761 | 4.0 | 41 | 3.1 |
| Other activities | 62,155 | 12.4 | 177 | 13.5 |

basic salary of a person living on mainland Ecuador.[11] This means that in the public sector the minimum wage on the Ecuadorian mainland is $292.00/month (in 2010), but in the Galapagos it is $584.00/month. Moreover, 10.7% of people work in the public sector, a much higher percentage than on the Ecuadorian mainland, where only 4.1% work in the public sector. Favorable economic conditions in the Galapagos Islands are present also in the private sector. According to the Life Condition Survey, conducted in 2009, the average monthly income for public and private workers in Galapagos was $772.03/month, whereas on the Ecuadorian mainland it was $251.70/month.

The structure of the labor market in the Galapagos is strongly related to tourist activities. Table 4.5 shows the structure of the market and the importance of tourism relative to other typical jobs in wholesale and retail, hosting activities and food services, transport and storage, and construction.

The market has benefited from the quantity of business per 1000 inhabitants. According to the last economic census (2010), in Galapagos there were 52.9 businesses per 1000 inhabitants, while on the mainland of Ecuador there were 35.3 businesses per 1000 inhabitants. Table 4.6 shows the number of businesses by economic activity sector.

---

[11] According to the Special Law for the Conservation and Sustainable Development of the Province of Galapagos issued on 1998 and reformed on 2003, "The minimum wage, minimum sectorial or basic wage of the province of Galapagos in each category consist of the sum of the minimum basic wage and minimum sectorial wage salary in continent plus 100% increase."

**Table 4.7** Access to information and communication technologies (INEC, Population and housing census, 2010a)

| ICT access | Ecuador (%) | Galapagos (%) |
|---|---|---|
| Conventional telephone | 33.4 | 68.7 |
| Cell phone | 76.3 | 92.1 |
| Internet | 13.0 | 18.3 |
| Computer | 26.3 | 46.4 |
| Pay TV | 17.5 | 33.2 |

**Table 4.8** Poverty disparities: mainland Ecuador and the Galapagos Islands (INEC, Population and housing census, 2010a)

| Poverty | Ecuador (%) | Galapagos (%) |
|---|---|---|
| Unsatisfied basic needs, poverty (households) | 56.2 | 47.6 |

A healthy labor market generates better connectivity (in terms of communications) and reduces poverty. In the Galapagos, there are proportionally fewer people living in poverty and very high communications connectivity. Compared to the Ecuadorian mainland, Galapagos is the best place for connections for households and businesses: 92.11% of households have at least one active cell phone, whereas 76.28% of households on the Ecuadorian mainland have at least one active cell phone; in the Galapagos 18.33% of households have Internet access, while on the Ecuadorian mainland 13.33% of businesses have Internet access. According to the 2010 National Economic Census, 21.9% of businesses in Galapagos use the Internet in their activities, while on the Ecuadorian mainland the rate is 11% (Table 4.7). Galapagos has fewer people living in poverty than does mainland Ecuador (Table 4.8).

## Unfavorable Socioeconomic Conditions of Galapagos

Unfortunately, the socioeconomic scenario is not perfect for residents of the Galapagos. The healthy labor market with high incomes also causes high commodity prices and, in the Galapagos, there are severe problems with access to basic services. The classic statistic to evaluate the average cost of life is the "cost of basic basket," i.e., goods and services needed to satisfy basic needs, which is composed of 75 fundamental goods for a typical family life. Table 4.9 shows the difference between the "basket" in the Galapagos and the same products on the Ecuadorian mainland.

In the Galapagos Islands, higher education levels occur as compared to mainland Ecuador,[12] although there are severe problems with education costs and infrastructure.

---

[12] According to the 2010 census, Galapagos has a lower illiteracy rate (1.3%) than mainland Ecuador (6.8%). The average number of years spent in school is 11.9 years in the Galapagos, whereas on the Ecuadorian mainland the average is reported to be 9.6 years.

**Table 4.9** Average "cost of basic basket" in US dollars (INEC, Consumer price index for Galapagos, 2010c)

| Date | Ecuador ($) | Galapagos ($) | Difference ($) |
|---|---|---|---|
| April 2009 | 522.76 | 835.32 | 312.56 |
| May 2009 | 522.75 | 839.61 | 316.86 |
| June 2009 | 522.38 | 843.00 | 320.62 |
| July 2009 | 521.73 | 844.86 | 323.13 |
| August 2009 | 519.30 | 847.30 | 328.00 |
| September 2009 | 521.26 | 841.30 | 320.04 |
| October 2009 | 522.34 | 861.60 | 339.26 |
| November 2009 | 522.59 | 860.83 | 338.24 |
| December 2009 | 528.90 | 862.64 | 333.74 |
| January 2010 | 534.33 | 865.11 | 330.78 |
| February 2010 | 535.48 | 868.74 | 333.26 |
| March 2010 | 535.56 | 868.98 | 333.42 |

**Table 4.10** Average cost of education in Ecuador and the Galapagos Islands (INEC, Living conditions survey, 2009a)

| Educational expenses | Ecuador (2005–2006) ($) | Galapagos (2009) ($) |
|---|---|---|
| Enrollment | 61.95 | 256.47 |
| Uniforms | 36.10 | 64.74 |
| Textbooks and school supplies | 48.47 | 68.50 |
| Monthly tuition | 45.28 | 72.87 |
| School materials | 7.73 | 13.29 |
| School transport | 19.74 | 21.08 |
| Others | 6.69 | 14.45 |

Using the 2009 Life Condition Survey of Galapagos, we can compare differences between educational expenses in the Galapagos and on the Ecuadorian mainland. In some cases, expenditures are more than 200% higher in the Galapagos as compared to the mainland (Table 4.10).

It is important to note that in Galapagos, one of every ten students (10%) is enrolled in distance learning, while on the Ecuadorian mainland, this proportion is minimal (1.1%). This phenomenon is likely caused by the relatively low educational opportunities in the Galapagos, especially for higher education (Table 4.11). There are no main campuses, only extensions that have their head-quarters on the mainland.

Galapagos presents a difficult situation in terms of basic services. Despite the physical limitations of the islands, during the last 10 years, the Galapagos experienced an increase in the number of housing units (68.7%), which is twice the national increase of 34.7%. In terms of basic services, Galapagos showed significant deficits, particularly in the potable water network coverage and sewer service: in 2001 only 30.8% of homes had network sewer service and by 2010 this proportion was reduced to 26.8%. On the Ecuadorian mainland, 48% of

**Table 4.11**  Types of educational establishments (INEC, National economic census, 2010b)

| Activity of the establishment | Ecuador | | Galapagos | |
|---|---|---|---|---|
| | Total establish. | Establish. per 1,000 inhabitants | Total establish. | Establish. per 1,000 inhabitants |
| Preprimary and primary education | 8,144 | 0.56 | 13 | 0.52 |
| Secondary general education | 1,903 | 0.13 | 9 | 0.36 |
| Technical and professional education | 497 | 0.03 | – | 0.00 |
| Undergraduate education | 547 | 0.04 | 4 | 0.16 |
| Sports and recreation education | 305 | 0.02 | 1 | 0.04 |
| Cultural education | 454 | 0.03 | – | 0.00 |
| Other education | 1,048 | 0.07 | 2 | 0.08 |
| Support activities for education | 183 | 0.01 | 1 | 0.04 |

homes had network sewer service in 2001 and 53.6% in 2010. These conditions are even worse on Santa Cruz, an island that has 61.3% of the total population of the archipelago, with a sewage system that only includes 3.5% of the households.

Finally, it is necessary to analyze the situation of women in the Galapagos: women in 2010 accounted for 42.8% of the total population, and illiteracy among women was 1.6% compared to men living in the Galapagos (0.6%) and women on the Ecuadorian mainland (7.7%).

In 2010, the average length of time women in Galapagos had gone to school was 12.1 years, which was greater than men (11.7 years) and substantially higher than the average for women on the Ecuadorian mainland (9.5 years). In the two previous censuses, the average time spent in school in the Galapagos was lower (8.13 years in 2001 and 8.32 years in 1990). Table 4.12 compares the education level of women in the Galapagos and on mainland Ecuador.

Women's increased access to education is reflected in their participation in the labor market. In 2010, the "economically active population" in the Galapagos was 12,975, of which 52.1% was composed of women, compared to 38.44% on the Ecuadorian mainland. According to the National Economic Census (2010b), 50.3% of establishments in Galapagos are owned or managed by women, as compared to 48.4% on the Ecuadorian mainland (Table 4.13).

Despite progress in reducing gender inequality in Galapagos, there is a disturbing presence of violence against women in the islands: 55.3% of women have suffered some kind of violence, and 43.3% have been victims of violence in a relationship. In terms of the type of violence, 35.3% of women have been physically abused; 49.9% suffered psychological violence; 22.8% reported being victims of

**Table 4.12** Education level of women in the Galapagos and on mainland Ecuador (INEC, Population and housing census, 2010a)

| Education level | Mainland (%) | Galapagos (%) |
|---|---|---|
| None | 5.6 | 1.5 |
| Literacy center | 1.0 | 0.4 |
| Preschool | 1.1 | 0.7 |
| Primary | 34.3 | 24.6 |
| Secondary | 22.9 | 24.2 |
| Basic education | 9.1 | 8.4 |
| Bachelor | 7.2 | 11.2 |
| Post bachelor cycle | 1.1 | 2.0 |
| Undergraduate | 14.2 | 19.6 |
| Postgraduate | 1.0 | 2.1 |
| Undeclared | 2.5 | 5.3 |
| Total | 100.0 | 100.0 |

**Table 4.13** Participation of women in the labor market (INEC, Population and housing census, 2010a)

| | 1990 | | 2001 | | 2010 | |
|---|---|---|---|---|---|---|
| Gender | Number | % | Number | % | Number | % |
| Men | 3,557 | 59.6 | 6,170 | 70.4 | 7,848 | 60.5 |
| Women | 2,412 | 40.4 | 2,598 | 29.6 | 5,127 | 39.5 |
| Total | 5,969 | 100.0 | 8,768 | 100.0 | 12,975 | 100.0 |

sexual violence; and 33.0% have suffered patrimonial or economic violence (National Institute of Statistics and Censuses 2011).[13]

# Health Situation Analysis

In general, people in the Galapagos are considered to be healthier than those on the Ecuadorian mainland, but there remains a limited availability of specialized health services on the islands to serve them. For many people, it is necessary to travel to the mainland (spending additional money and time) to resolve routine or complex health needs.

Using the 2009 Life Condition Survey of Galapagos, basic indicators relate to vaccines for children under 5 years old: Pentavac vaccine has 29.5% more coverage than on the Ecuadorian mainland; the SRP vaccine has 16.2% more coverage than on the Ecuadorian mainland; and chronic (height versus age), global (weight versus age), and acute (weight versus height) malnutrition have 6.3, 6.8, and 0.7%,

---

[13] National Institute of Statistics and Censuses (2011) National survey of family relationships and violence against women

respectively, less incidence in the Galapagos than on the Ecuadorian mainland (Table 4.14).

Another group being studied is the women of childbearing age (WCA). In this case, it is possible to identify more coverage in the Galapagos Islands for tetanus and rubella vaccinations, more *Papanicolaou* (Pap) exams, and more knowledge of family planning methods (Table 4.15).

According to the 2010 census, structural health indicators are relatively high in the Galapagos as compared to the Ecuadorian mainland. For instance, a greater proportion of people subscribe to social security, particularly in the private and public sectors (Table 4.16).

Additionally, there is more assistance for childcare systems in Galapagos (19.8%), as compared to the Ecuadorian mainland (13.20%). In particular, in Galapagos the private system of childcare seems to take the place of religious (church) and NGO childcare (Table 4.17).

Despite this positive scenario, there are deficiencies in responding to specialized needs and health emergencies in the Galapagos, as there is a general lack of infrastructure for treating infectious diseases and for practicing oncology, dermatology, pediatrics, traumatology, psychiatry, and other services (Table 4.18).

Health care in the Galapagos is challenged by the limited availability of specialized services and the general lack of a diverse health services infrastructure (Table 4.19).

Finally, we present statistics related to pregnancy, showing the care provided on the Ecuadorian mainland and in the Galapagos. Table 4.20 shows the proportion of births in private and public institutions. As we can see, in the Galapagos a high proportion of mothers travel to the mainland to give birth (33%). This is one of the most important statistics that reveals the incompleteness of the islands' health system and shows the importance of having public health systems.

The basic question to ask about the health situation in Galapagos is how sustainable is it? In a place where favorable health conditions exist, but where the demand for more specialized services is growing with the population, can services become compatible with the demands and needs of the population? Again, the intuitive solution would be to find a way to limit population growth, or the unsatisfied needs of the actual population will continue, especially for services related to aging.

## Conclusions

This chapter has summarized the demographic, socioeconomic, and public health situation in the Galapagos Islands, using the latest official Ecuadorian statistics. The data reveals the *Galapagos Paradox*: the Galapagos is a place with a healthy economy and good living conditions, but which has some major problems related to uncontrolled and unmeasured migration. While migration appears necessary for the tourism industry, which is very important for the economy in the Galapagos, the direct and indirect effects of the population increase will lead to severe problems in

**Table 4.14** Vaccine coverage for children under 5 years old (INEC, Living conditions survey, 2009a)

|  | Mainland (2005–2006) (%) | Galapagos (2009) (%) |
|---|---|---|
| BCG vaccine | 98.3 | 99.0 |
| Pentavalent vaccine | 65.5 | 95.0 |
| Polio vaccine | 93.6 | 95.0 |
| SRP vaccine | 62.9 | 79.1 |
| Chronic malnutrition | 18.1 | 11.8 |
| Global malnutrition | 8.6 | 1.8 |
| Acute malnutrition | 1.7 | 1.0 |
| Diarrhea presence | 25.0 | 8.3 |
| Respiratory diseases | 56.0 | 45.3 |

**Table 4.15** Vaccine and health prevention coverage in women of childbearing age, i.e., between 12 and 49 years old (INEC, Living conditions survey, 2005–2006, 2009a)

|  | Ecuador (2005–2006) (%) | Galapagos (2009) (%) |
|---|---|---|
| Tetanus vaccine | 86.1 | 96.0 |
| Rubella vaccine | 74.7 | 82.8 |
| Papanicolaou examination | 51.3 | 72.9 |
| Knowledge about family planning | 92.5 | 5.6 |

**Table 4.16** Social security coverage on the Ecuadorian mainland and Galapagos (INEC, Population and housing census, 2010a)

|  | Ecuador (%) | Galapagos (%) |
|---|---|---|
| Public social security | 27.2 | 37.4 |
| Private social security | 9.4 | 17.9 |

**Table 4.17** Type of institution providing childcare (INEC, Living conditions survey, 2006–2006, 2009a)

|  | Ecuador (2005–2006) (%) | Galapagos (2009) (%) |
|---|---|---|
| Public | 78.1 | 76.2 |
| Private | 13.4 | 23.8 |
| Church/NGO's | 8.5 | 0.0 |

the future, with regard to people's well-being and quality of life. In the last two decades, the growth of the migrant population has been aided by the presence of a poor registration system and a too informal process of granting residence cards. The sustainability of the system depends on controlling these problems.

**Table 4.18** Morbidity by type of establishment (INEC, Annual hospital discharge, 2009b)

| Establishment | Ecuador | | Galapagos | |
|---|---|---|---|---|
| | Frequency | % | Frequency | % |
| Total | 1,090,263 | 100.00 | 1,938 | 100.00 |
| Basic hospital | 177,977 | 16.32 | 858 | 44.27 |
| General hospital | 370,238 | 33.96 | 797 | 41.12 |
| General clinic (no specialized) (private) | 237,493 | 21.78 | 126 | 6.50 |
| Pediatric hospital | 66,961 | 6.14 | 71 | 3.66 |
| Specialized hospital | 106,262 | 9.75 | 44 | 2.27 |
| Obstetrics and gynecology hospital | 87,369 | 8.01 | 20 | 1.03 |
| Cancer hospital | 21,239 | 1.95 | 9 | 0.46 |
| Hospital for infectious diseases | 2,824 | 0.26 | 7 | 0.36 |
| Psychiatric hospital and sanatorium of alcoholics | 3,123 | 0.29 | 3 | 0.15 |
| Obstetrics and gynecology clinic | 11,298 | 1.04 | 2 | 0.10 |
| Pneumological hospital | 2,177 | 0.20 | 1 | 0.05 |
| Dermatological hospital | 298 | 0.03 | – | – |
| Geriatric hospital | 1,954 | 0.18 | – | – |
| Pediatric clinic | 120 | 0.01 | – | – |
| Trauma clinic | 237 | 0.02 | – | – |
| Psychiatry clinic | 48 | 0.00 | – | – |
| Other specialized clinics | 645 | 0.06 | – | – |

**Table 4.19** Number of beds available by type of establishment (Yearbook of hospital beds, 2010e)

| Establishment | Ecuador | | Galapagos | |
|---|---|---|---|---|
| | Total available beds | % | Total available beds | % |
| Ministry of Public Health | 8,484 | 35.67 | 30 | 100.00 |
| Ministry of Justice and Police and Government | 247 | 1.04 | – | – |
| Ministry of National Defense | 709 | 2.98 | – | – |
| Social Security Institute | 2,143 | 9.01 | – | – |
| Other publics | 160 | 0.67 | – | – |
| Municipalities | 170 | 0.71 | – | – |
| Universities and polytechnics | 146 | 0.61 | – | – |
| Charity Board of Guayaquil | 2,496 | 10.49 | – | – |
| Society Against Cancer | 560 | 2.35 | – | – |
| Fisco Misionales | 88 | 0.37 | – | – |
| Private nonprofit | 714 | 3.00 | – | – |
| Private for-profit | 7,867 | 33.08 | – | – |
| Total | 23,784 | 100.00 | 30 | 100.00 |

**Table 4.20** Point of care in birth assistance (INEC, Living conditions survey, 2006–2006, 2009a)

| Point of care in the last birth assistance | Ecuador (2005–2006) | | Galapagos (2009) | |
|---|---|---|---|---|
| | Frequency | % | Frequency | % |
| Hospital MSP[a] | 603,424 | 44.30 | 1,394 | 68.00 |
| Health center MSP | 24,570 | 1.80 | 4 | 0.20 |
| Health subcenter MSP | 15,214 | 1.10 | 31 | 1.50 |
| Hospital IESS | 54,896 | 4.00 | 15 | 0.70 |
| Health center IESS | 3,948 | 0.30 | 0 | 0.00 |
| Health subcenter IESS[b] | 1,801 | 0.10 | 42 | 2.00 |
| Hospital/PSJ/FFAA/ISSPOL[c] | – | – | 60 | 2.90 |
| Private hospital or private clinic | 396,599 | 29.10 | 246 | 12.00 |
| Private health center | 21,537 | 1.60 | 123 | 6.00 |
| Private practice | 29,587 | 2.20 | 93 | 4.60 |
| House midwife | 11,367 | 0.80 | 0 | 0.00 |
| Home | 196,519 | 14.40 | 40 | 1.90 |
| Other | 1,362 | 0.10 | 0 | 0.00 |

[a]Ministry of Public Health
[b]Social Security Institute
[c]Police and Army Health and Social Security

People on the islands enjoy a good economic situation overall but, at the same time, they suffer from high prices, poor access to basic services, and deficiencies in health care and educational infrastructure. Galapagos has become an ideal setting for short-term migrants; a place to obtain money, but also a place to leave once that objective is achieved.

It is urgently necessary to have a public policy intervention in the islands. In this document, we discuss the context, but not the kind of policies that will be required to stem the flow of people and to create a sustainable social–ecological environment for the islands. It is necessary to decide upon one of two future paths: continue tourism growth and adapt to its consequences, or define specific limits for economic activities and rethink the living conditions of the actual habitants as well as tourists.

Further analysis should include specific metrics and sustainability models for the Galapagos to generate a public policy discourse about the future vision of the islands. We strongly recommend a multi-criteria analysis as a way to synthesize ecological and socioeconomic problems in a sensitive area without weighing or prioritizing any single dimension of development.

# References

Galapagos National Park, Ministry of Environment (2011) Annual report of Galapagos immigration. Galapagos, Ecuador
National Congress (1998) Special law for Galapagos province. Quito, Ecuador

National Institute of Statistics and Censuses (2005–2006) Living conditions survey
National Institute of Statistics and Censuses (2009a) Living conditions survey of Galapagos
National Institute of Statistics and Censuses (2009b) Annual hospital discharge
National Institute of Statistics and Censuses (2010) Population and housing census
National Institute of Statistics and Censuses (2010) National economic census
National Institute of Statistics and Censuses (2010) Consumer price index for Galapagos
National Institute of Statistics and Censuses (2010) Yearbook of vital statistics
National Institute of Statistics and Censuses (2010) Yearbook of hospital beds
National Institute of Statistics and Censuses (2011) National survey of family relationships and
    violence against women
Tapia W, Ospina P, Quiroga D, Gonzalez JA, Montes C (2009) Science for Galapagos: a proposed
    strategy and priority research agenda for sustainability of the archipelago. Quito, Ecuador
Vanguardia (2012) The Galapagos report. http://www.revistavanguardia.com/index.php?option
    =com_content&view=category&layout=blog&id=46&Itemid=106. Accessed Apr 2012
World Commission on Environment and Development (1987) Report of the eighth and final
    meeting of the World Commission on Environment and Development. Tokyo, Japan
World Wildlife Fund (2003) Migration and environment in the Galapagos Islands. Quito, Ecuador

# Chapter 5
# Environmental Crisis and the Production of Alternatives: Conservation Practice(s) in the Galapagos Islands

Wendy Wolford, Flora Lu, and Gabriela Valdivia

## Introduction: Environmental Crisis and Conservation in the Galapagos Islands

Few causes seem to mobilize support today like biodiversity conservation. The United Nations named 2010 as the International Year of Biodiversity in recognition of the rapid rate of species extinction. Monitoring programs such as the online Encyclopedia of Life (EOL), founded by biologist E. O. Wilson, represent efforts across the globe to provide a digital compendium of conservation that accumulates and makes accessible scientific knowledge about all existing species and identifies those most at risk. And concern over biodiversity as a resource at risk—ecologically and economically vital, finite, and threatened by human activities—has also led to an explosion of environmental advocacy institutions, governmental and nongovernmental organizations (NGOs), and programs regulating the effects of human activity on nonhuman species (Sodikoff 2012).

Thus, it is not surprising that recent accounts on the state of conservation in the Galapagos Islands have stirred heated discussions about the future of conservation practice. For example, in a recent article in *Science* titled "Embracing Invasives" (Vince 2011), Mark Gardener, head of the Division of Terrestrial Science at the

W. Wolford (✉)
Department of Development Sociology, Cornell University,
Ithaca, NY 14583, USA
e-mail: www43@cornell.edu

F. Lu
Department of Environmental Studies, University of California,
Santa Cruz, CA, USA

G. Valdivia
Department of Geography, University of North Carolina,
Chapel Hill, NC, USA

S.J. Walsh and C.F. Mena (eds.), *Science and Conservation in the Galapagos Islands:* *Frameworks & Perspectives*, Social and Ecological Interactions in the Galapagos Islands 1, DOI 10.1007/978-1-4614-5794-7_5, © Springer Science+Business Media, LLC 2013

Charles Darwin Research Station (CDRS), the longest established research facility on the islands, is quoted as saying that "it's time to embrace the aliens" and that conservationists need to recognize the futility of chasing "original" landscapes and "optimize these new ecosystems." Vince's article generated a backlash of criticism against both Mark Gardener and the idea of accepting invasive species. As Vince herself suggests in the *Science* piece:

> ...Gardener's decision to abandon the fight to preserve and restore indigenous-only species here has caused shock waves among the venerable members of the Charles Darwin Foundation, the 50-year-old organization that runs CDRS, with many of the old guard "very upset by the idea," Gardener says. William Laurance, a conservation ecologist at James Cook University in Cairns, Australia, is also concerned: "If people want to resign themselves to managing novel ecosystems—and it sounds like that's the reality they face on the Galapagos—then what we're doing is homogenizing the world's biota; setting the world on a geological epoch: the Homogocene."

*Science*'s article and the ensuing debate highlight how concerns over change, continuity, and crisis dominate debates about the state of conservation and the need for action in Galapagos—a moral imperative. Conceptually, these "moral geographies" (Bryant 2001), or the spatial envisioning of what are believed to be proper ways of knowing, regulating, and acting upon how humans relate to nature, shape conservation decisions and effects. Solutions to biodiversity loss often lead with notions of scarcity and loss that are treated as objective quantitative categories, when they are actually normative qualifications that award moral standing and relative value within specific historical and geographical contexts (Escobar 1995; Leopold 1949; Neumann 2004; Peet and Watts 2006).

In Galapagos, a more nuanced approach to environmental crisis is necessary. The *Science* article cited above repeats a well-known rendering of the proper relationship between humans and nature, one where introduced species should be rejected, pristine landscapes protected, and original species or systems kept free of humans (Agrawal and Sawyer 2001; Cronon 1995; Holt 2005). This rendering ignores context, history, and social needs, all of which shape the social construction of biodiversity conservation and the impossibilities of making a purely scientific evaluation (Braun 1997; West 2006). As of 2011, 25,000 people live on the islands, due to a combination of homesteading programs in the 1950s and the attractiveness of the more recent, burgeoning tourism industry. Regional institutions and conservationists have presented conservation as an expost facto attempt to limit the presence of human residents (Quiroga 2009), while Galapagos residents insist that they have rights to the islands and could be co-caretakers of well-managed, biodiverse landscapes.

A closer look at the conservation landscape in Galapagos suggests that there is much more to the current story than a debate about whether conservation of original landscapes and species has reached a crisis or not. In this chapter, we offer an alternative reading of the state of conservation in Galapagos, based upon research on the terrestrial areas of the archipelago. Drawing on political ecology insights, we propose that conservation practices have multiplied and adapted to diverse locations, specializing in particular sites such that there is not *one* approach to conservation but several, all of which sit in uneasy and unequal but productive tension together. These approaches incorporate perspectives from a diversity of groups, partnerships, and interests within the islands as well as with mainland Ecuador and beyond (Ospina and Falconi 2007). From eradication programs and policing of boundaries between people and protected areas to the recognition of multifunctional landscapes, conservation has taken many forms and engaged distinct views, interests, and actors. This somewhat chaotic, contested cobbling of approaches to conservation is not, however, indicative of failure (cf. Simberloff et al. 2011); rather, it is the product of negotiation, resistance, dispute, and accommodation between local residents, resource users, scientists, and park officials, some of whom now publicly acknowledge the need for a conservation science that does not separate humans from nature in an attempt to preserve valuable ecosystems (Gonzalez et al. 2008). Just as in nature, there may be value in diversity for conservation policy and science as well.

We make two additional arguments about this multiplicity of conservation. First, the diversity in conservation approaches is a product of struggles over governance. Clashes between those who have different visions of conservation and development have historically produced periods of extreme tension. For example, in demanding better access to marine resources, fishermen have gone on strike: in 1994, the national park offices were invaded and vehicles lit on fire; in 1997, strikes again threatened governance; in 1999, the house of the director of the Galapagos National Park Service (GNPS) was torched; and in 2001, violent confrontations took place between residents and park representatives. These moments do not represent either scientific disputes or "riots of the belly" where people protest conditions of bare necessity (Thompson 1971). By almost any measure, life in Galapagos is considerably easier, more secure, and better off than life on the mainland. Rather, the conflicts point to the struggles and negotiations between different views on the ideal use of and access to resources. As political ecologists suggest, labeling these at-times-violent struggles as *the* problem does not offer a resolution to crisis. Instead, these moments should be examined as ways in which aspects of political life are taken up and reconfigured through environmental claims; struggles over environment are simultaneously struggles over social identity, belonging and exclusion, and rights (for example, see Peet, Robbins and Watts 2011).

Second, conflicts provide the potential for resolution. Moments of extreme tension—what are dubbed "crises" locally and in the popular press—become potentially generative times and spaces in which new attitudes, alliances, resources, and approaches have been discovered and partnerships made. It is during the aftermath of "crisis" that negotiations between different interest groups have been most evident. And each negotiation has its particular spatial and social characteristics, representing distinct ethical complexities.

This chapter draws on field research conducted during the summers of 2007, 2008, 2009, and 2011 to elaborate on the state of conservation practice in Galapagos. The coauthors interviewed 105 local residents, including farmers, fishermen, tourism providers, municipal leaders and administrators, employees and officials with the Galapagos National Park, local organizations, and conservationists with different agencies, including the Charles Darwin Research Station.

These interviews were transcribed, written out, and analyzed for dominant themes and perspectives. All direct quotes are presented anonymously to protect the identities of research participants as promised. In the case of quotes from highly visible figures, such as park officials and local politicians, we indicate the use of real names. In addition to the interviews, we attended meetings, visited farms, analyzed farmers' markets, and observed people at work in various occupations.

The next section of this chapter presents our theoretical framework for thinking through crisis as both constituting and constitutive of change. We discuss the specific role that crisis has played in creating new spaces for conservation during the brief history of the Galapagos National Park, which celebrated its 50th anniversary in 2009. We then elaborate on what those spaces look like by discussing four different conservation approaches at work in the islands today. The four projects include: (1) the project to eradicate goats introduced in large numbers on the island of Isabela, which was widely hailed as a successful one that established a clear separation between the park and human-occupied areas; (2) the Galapagos National Park's Plan for Total Control, which focuses on monitoring the spread of invasives across the border between the protected areas of the reserve and the inhabited farmland; (3) the project from the Charles Darwin Research Station to calculate the human footprints of different actors or groups across the islands, which represented the station's efforts to become more involved with social issues; and (4) projects promoted by both the municipalities of San Cristobal and Santa Cruz as well as a local nongovernmental organization called FUNDAR (Foundation for Alternative Responsible Development) geared toward increasing organic and agroecological farming practices that together would constitute "working landscapes" along the border between the agricultural areas and the park.

We use these projects as windows onto the diversity of conservation as it is taken up in distinct sites, rather than as an attempt to describe or represent conservation in its entirety.

## Background: The Galapagos as a Case Study of Conservation

The Galapagos Islands are widely known for their biological uniqueness and natural beauty. Free of humans and predators for most of their history, these "enchanted islands" have developed some of the most unique life-forms on the planet, highly adapted to their harsh surroundings and living in ecological isolation. It was not until Charles Darwin's famous visit in 1835, however (a visit which helped inspire the theory of evolution by natural selection), that this archipelago began to receive international recognition. In 1959, the Galapagos National Park was formed, and in 1973, the archipelago was incorporated as the twenty-second province of Ecuador.[1]

---

[1] State-government representation in the archipelago includes rural associations, municipal governments, and governorship.

UNESCO designated the Galapagos as a World Natural Heritage Site in 1978 to honor the "outstanding universal value" of the "magnificent and unique" natural features of the islands and to ensure their conservation for future generations.

Over the past three decades, dramatic changes have occurred in the terrestrial and marine ecosystems of the Galapagos. As a result of the international recognition and popularity of their unusual and endemic species (e.g., giant tortoises, marine iguanas, and ground finches), the Galapagos Islands have become home to a rapidly growing ecotourism industry. In 1990, the number of visitors to the islands was 40,000, and by 2010, the number had increased to 190,000. Since the 1970s, the islands have also drawn thousands of new residents attracted by the promise of lucrative opportunities linked to construction and tourism. From 1990 to 2001, Galapagos province had the highest population growth rate in the country at approximately 6%. For these new residents, the promise of profits was a welcome change from economic crisis, social upheaval, and political volatility on the mainland.[2]

By 1999, Ecuador's GDP was nearly equal to its debt load (at $13.75 billion), poverty was at 40%, and nationwide unemployment increased to 15% (Jokisch and Pribilsky 2002, p. 76), considerably higher than unemployment in the Galapagos Islands (Ospina 2006). Increasing numbers of Ecuadorians moved into the coastal communities and highland agricultural zones that comprise the 3% of the archipelago that is available for habitation (Boersma et al. 2005). In a place valued by many for its unique landscapes and biodiversity, demographic growth and economic development are seen as "invasive" or as resulting in the spread of unwanted species (introduced flora and fauna such as blackberry, guava, and goats). Concerns about the spread of invasives and fear for the survival of native species have historically led to fortress conservation policies that pit local inhabitants—Galapagueños—against GNP authorities and conservation scientists (Macdonald 1997). Farmers and fishermen argue that they, as residents of the islands, have rights to the resources. But conservation scientists affiliated with the CDRS and World Wildlife Fund and employees with the GNP have argued that more stringent regulations and effective sanctions are necessary because the growth of the local population and local economies—associated with the growth in tourism and fisheries—leads to unprecedented overharvesting of resources, pollution, habitat change, and introduction of invasive species (Watkins and Cruz 2007).

These kinds of socio-environmental conflicts have reshaped the nature of the debate on the islands, in part because the conflicts themselves have led to the production of new laws to regulate human–society relations in the Galapagos. The increase in population prompted a UN investigation in 1996, upon which the

---

[2] Ecuador fought a costly border war with Peru in 1995 and bled another US$2 billion in economic damages from El Niño floods in 1997–1998, which crippled banana exports and infrastructure. In addition, the price for petroleum, Ecuador's most lucrative export, fell to a near record low about that time. In early 1999, then-president Jamil Mahuad consolidated, closed or bailed out 16 financial institutions during a banking crisis, and antagonized the citizenry by freezing the majority of bank accounts (in an effort to stop capital flight) and agreeing to dollarize the economy as a concession to the IMF (the sucre as a result was devalued 66%).

islands almost lost their World Heritage status, but the Ecuadorian government enacted special legislation—the *Special Law for Galapagos*—to more tightly control human migration from the mainland and the introduction of invasive species to Galapagos. As a recent director of the Charles Darwin Research Center said, the Special Law was "more of a vision than a law"; it attempted to resolve the growing tensions between conservation and development by restricting migration to the islands, fortifying the existing institutional structure, and implementing new channels for participatory management. Under the umbrella of the Special Law, the Galapagos Marine Reserve was created in 2001 to regulate the extraction of marine resources in the islands. The Special Law provides a legal comanagement framework through which state institutions, the GNP, and local actors negotiate conflicting interests over marine resources. The law also provided more resources for the island residents and administrators, as all non-Galapagos visitors were subsequently required to pay a US$100 entrance fee. This money is divided between the Galapagos National Park (45%), the municipalities of each island (25%), the town mayors (10%), INGALA (10%), SIGAL (5%), and the armed forces (5%). These actions were received favorably by UNESCO, and the committee agreed not to revoke World Heritage status. Between 1998 and 1999, UNESCO approved over US$4 million in funding for the park.

Less than a decade later, however, the islands were in trouble again. On April 10, 2007, the Galapagos Islands were officially declared to be "at risk" and UNESCO placed the archipelago on its list of World Heritage Sites in danger. Ecuador's President, Rafael Correa, publicly decried the "institutional, environmental, and social crisis" that plagued the islands and declared that conservation would become a "national priority."[3] We elaborate more on the nature and result of this crisis in the sections that follow.

## Pulp Fictions of Conservation: A Theoretical Framework

In her analysis of indigeneity in Brazil (1998), anthropologist Alcida Ramos draws on the term "pulp fictions" to elaborate on the complex ways in which identifications are negotiated even as unequal power relations shape the terrain upon which representations do their work. In Ramos' case, superstitions, myths, and romantic idealizations of indigenous peoples in Brazil represent collectively held beliefs about the "proper" and moral relationship between different categories of humans and nature. The belief that indigenous peoples are natural stewards of the land is a discourse that is widely accepted by those who equate indigenous peoples with "raw nature." The underside of this belief is that indigenous peoples are "wild" and untamed, not responsible for what they do and not entirely capable of negotiating

---

[3] For news reports of President Correa's statement, see http://news.bbc.co.uk/2/hi/americas/6543653. stm.

the modern world. Although this discourse is essentializing, it can be strategic; indigenous groups also appropriate the narratives of "raw nature" as a tool in the fight to control territory. How the discourse works—or what work it performs—depends on the sociopolitical context and the relations between different actors engaged in its production and consumption.

We borrow this conceptual strategy to elaborate a similar argument about conservation in Galapagos: conservation policies inherently represent a perspective on the appropriate relationship between people and the environment. These perspectives are fueled by different interests that struggle for physical and symbolic space such that what appear to be clear and straightforward "problems" (e.g., goats and blackberry take over landscapes and thus must be eradicated) are actually intensely disputed renderings of socio-natural relationships. Different institutions and groups have different perspectives on the relationship between humans and nature, and so conservation necessarily has multiple meanings. Just as the myth of the noble savage became something different for different groups in the Brazilian Amazon, solutions to resolve the environmental crisis in the Galapagos are adapted and modified through their engagements with local scientists, managers, residents, and target species. Mixed results or unexpected developments remind us that there is no guarantee of the appropriateness of one conservation approach over others in such a diverse and dynamic archipelago (Atkinson et al. 2008; Gardener et al. 2010b). As declarations of ecological crises become more frequent in light of contemporary threats such as climate change and biodiversity loss, our goal is to contribute to the development of approaches that facilitate comparative analysis and a better understanding of crisis as a discourse, a space, and a site for the negotiation of new positions, identities, and frameworks for governance.

Political and human ecologists have highlighted the subjective nature of crises such as natural hazards: the experience and evaluation of any given "crisis event" depends in part on a person or group's relative exposure to risk (Blaikie et al 1994; Pelling 2003). Ultimately, it is clear that a crisis does not exist objectively, independent of humans: a crisis is a relationship between humans and their environment and between individuals with differential political power. With all of the rich studies investigating the nature and causes of environmental crises, there has been less systematic research on crises as a set of productive processes. Environmental "crises" are moments of conflict around human–environment relations that demand urgent action for a resolution. In response, the language, methods, and strategies of quantifiable conservation science are commonly used to frame the need for disciplining an "unruly terrain" that requires management and intervention (Crush 1995). A dominant discourse emerges which presents the causes, consequences, and correctives for the crisis. Although most media reports and official communications reiterate this discourse—through representations of loss, chaos, and devastation (cf. Bassett and Zuéli 2000)—there are, of course, other competing understandings of the problems and solutions associated with the crisis situation. As in the Galapagos, these views are based in differing ways of knowing and perceiving the situation, shaped by a multitude of interrelated factors, including people's relationship to the resource base and their political, economic, and sociocultural position

(Lu, Valdivia and Wolford forthcoming). Thus, crisis should not be taken as given but deconstructed to enable the incorporation of different possible interpretations and experiences.

Oftentimes, crises are formalized through a commonsensical discourse of "better governance" (the diversity of opinions notwithstanding) based on the identification of a suitable knowledge base and definition of the dynamics of causality, effects, and remedy; the next step is the management phase. Political maneuverings result in laws and policies that start to effect tangible and concrete changes and impacts in people's lives and the landscape. Environmental management practices that draw on notions of resource scarcity often see crises emerging from institutional failures, that is, from breakdowns associated with the regulation of society and territory, such as tenure insecurity, weak political institutions leading to open access, and inability to achieve the collective action needed for conservation (Hardin 1968; Ostrom and Nagendra 2006; Guyer and Peters 1987; Turner 1999). It is often in moments of crisis that new spatialities of management—or conservation territories—are delineated, organized, and regulated in an effort to govern human–environment relations. In what follows, we show how ongoing crises in the Galapagos have been translated into mandates for action and better governance.

## The Production of Policy: Four Different Attempts to Manage Ecological Crisis

The Special Law of 1998 brought more money to the park and local residents and, according to park administrators, was responsible for enabling the restructuring of the park administration and training of employees and administrators. The ongoing crisis on the islands was explained to us variably in 2007 and 2008 as a "perfect storm" of machinations on the mainland, the chaos of the short-lived presidency of Gutierrez, institutional overload, rising demand in the form of tourism, and demographic pressure (see Lu, Valdivia and Wolford forthcoming).

The UNESCO designation of the islands as a World Heritage Site "in danger" came on the heels of unrest already occurring on the islands and local attempts to refashion the primary institution on the islands responsible for administration and oversight: the Galapagos National Park Service (GNPS). One GNP manager argued that the new management plan for the GNPS came out of a "terrible moment of crisis" that began as early as 2003. The GNPS had historically been the focal point for tensions on the islands as local residents argued that the park service was overly punitive in restricting access to living space and natural resources (in our interviews with local residents, this complaint was still very evident). In an attempt to attend to these tensions and negative perceptions, the GNPS organized community meetings to discuss its own structure and potential reorganization. The meetings brought together local residents and officials; a total of over 400 people met between 2004 and 2006. While the focus of the GNPS remains the conservation of the "indigenous environment of the islands," many of its policies emphasize sustainable livelihoods and conservation.

When Rafael Correa was elected president in 2006, his call for a new constitution provided further opportunity for the restructuring of the GNP. According to interviewees, new components of the park management plan outline the need for participatory conservation methods as well as a more technical section that provides the institutional support for park guards and managers to receive training or professionalization so that the GNP no longer has to depend on external institutions such as the Charles Darwin Research Station for scientific guidance; as we elaborate below, CDRS is the research arm of the Belgian-based Charles Darwin Foundation (CDF). As a park service administrator said, "before, we used to monitor penguins just because we were told to monitor them and now we monitor them because we know they're an important part of the ecosystem."

The park service still does not have enough staff, which forces employees to negotiate conservation priorities on the ground "as they go." The head of park management emphasized the park's "mistica de trabajo" (work culture) with single individuals in charge of multiple areas and not enough *piernas* (legs, or people). He expressed hope for the implementation of the new management plan of 2005, which provides the conceptual and technical tools to create more partnerships between park and people and local institutions. These partnerships do not constitute formal targets, as they might have in the past, rather the new plan privileges process over specific deliverables; the plan "...doesn't have deadlines, we will construct the plan as we go." The focus on process complements a parallel move away from species' specific conservation efforts to more ecosystem management. Interviewees suggested that a growing number of park employees believed that the time had come to focus on restoration and control rather than eradication: "time to start putting things in [not just tearing them out]." A former park director argued that the park could focus on eradication in other [uninhabited] islands but should strive for *control* in inhabited/large ones.

The focus on process, ecosystems, and control necessarily implies greater collaboration with local residents. The Special Law provided the *impetus* to form committees of farmers and park employees who would meet to discuss invasive species eradication and control on private land, but it is the new management plan that provides the institutional *tools* for doing so. Additionally, there is a normative shift as park employees increasingly recognize the value of participatory management, emphasizing that the park needs to be visible in the community. Organizers with local associations largely agree that the park is now working with formerly marginalized residents, such as farmers, and is more responsive although there are still "hard liners" who argue that the park should not be involved in "social" issues. Even leaders of the fishing cooperatives agree that participation could work. The president of one of the main fishing cooperatives on San Cristobal argued in 2009 that the new leaders of the park were more open to dialogue, and so the fishermen were trying to not strike or actively protest but were waiting to see whether collaboration would work. The president argued that the fishermen and the park were natural partners in conservation, but they needed to find approaches that would allow the fishermen to be "productive."

At the same time, the main scientific unit on the islands, the Charles Darwin Research Station, also saw the crisis of 2007 and the "at-risk" designation as a sign that more social science was necessary. The CDRS is the research branch of the Charles Darwin Foundation, a Belgian-based, international nonprofit organization founded in 1959. Its mission is to "provide knowledge and assistance through scientific research and complementary action to ensure the conservation of the environment and biodiversity in the Galapagos Archipelago."[4] To that end, the CDRS was created in 1964 and located on the main populated island of Santa Cruz. The station has approximately 120 affiliated staff and researchers who gather and disseminate scientific data on biodiversity, climate change, ecological restoration, and more. The station also operates several internationally famous tortoise breeding and repatriation programs. In 2008, over 270 scientists worked at the station in various temporary capacities. The station's activities are funded by governmental organizations (22% of total funding in 2007), sales of services and goods (20% of total funding in 2007), and private charitable donations (58% of total funding in 2007).[5] Until recently, it was widely argued that the station neglected study of the social or human environment. A former employee of the station said that the station studied and watched over the protected areas and that the problems in the social sector were seen as not as serious and knowledge of the underlying issues was idiosyncratic and anecdotal, not systematic.

In an attempt to negotiate the tension between conservation and development in response to the at-risk designation, there are now an increasing number of policies and programs intended to promote conservation. Some of these new programs rely on participation, and some continue the focus on territorial management, with strict separations between protected and residential zones. In what follows, we describe four different sites of conservation. Evolving over time, shaped by various moments of crisis, the four conservation sites are stitched together unequally, with vested interests supporting each one. The alliances that support each approach are in constant flux, as the interests and actors involved are negotiating, shifting, and making things up as they go—even as they work in a broader structural context that itself moves beneath their feet. For example, the notion of "working farm landscapes" supported by a local grassroots organization FUNDAR had very little space on the islands. The organization was sustained mostly by the enthusiasm and dedication of a small staff, and their work represented a significant divergence from other institutions that were geared more toward conservation than livelihoods. In 2009, however, the NGO received US$3 million in funding from the European Union, and the increased revenue plus the general shift within the park toward an acceptance of farmscapes as potential conservation landscapes has given the organization a more substantial profile on the islands. Increasingly, FUNDAR has gone from a relatively marginal and radical institution to one that actively collaborates with the park and the station on participatory projects such as household recycling and anti-dengue campaigns.

---

[4] See the CDRS website at http://www.darwinfoundation.org/english/pages/interna.php?txtCodiInfo=3.
[5] See the 2007 Annual report, p. 40, at http://www.darwinfoundation.org/english/_upload/anual_report_2007_1.pdf.

These three institutions—the GNPS, the CDRS, and FUNDAR—form part of the web of governance that runs through the following examples of attempts to negotiate the crisis of 2007.

## Invasive Species Eradication

Eradication refers to the elimination of every individual of a species from an area in which recolonization is unlikely to occur (Myers et al. 1998). The project to remove goats from the island of Isabela is a paradigmatic example of conservation as eradication.[6] This project, known as "the Isabela Project," began when funders of the Charles Darwin Foundation realized the extent of the goat problem on the largest inhabited island. Seen as a moral imperative to "save" nature, the Isabela Project articulates a clearly spatialized hierarchy of idealized positions, fixed in both time and place. Isabela is an island that conservationists value because it is still almost entirely "intact." The CDF began to focus attention on the goat problem in 1995, and in 1997, the Galapagos National Park together with the CDF held an international meeting with scientists who had worked in similar ecosystems. These scientists were recruited to help brainstorm ways of addressing goat eradication and ecosystem renewal. As one of the leaders of the project said, "We put out a call to the world and said, we're the Galapagos and we need help." Scientists responded, with over two dozen people in attendance from Australia, New Zealand, Argentina, the USA, Europe, and more. Several ideas were discussed and rejected. One idea was the classic biological approach of introducing plants with hormones that when ingested by the goats would cause sterility. This idea was rejected because, once sterile, the goats would still have many years ahead of them during which native plants would be eaten with voracity. Another idea floated was to bring in the Ecuadorian military to hunt the goats, but this was rejected because it was physically difficult to navigate the terrain and vegetation on the northern end of the island. It was also, as the former project leader cited above said, difficult to trust people you did not know because "who knows what they might do to a tortoise?" A third idea brought up by local residents was to hire local fishermen and hunters to kill the goats; this idea was favored by the residents because they could eat the goat meat and be paid for their labor, but it was rejected on the grounds that it would take too long and be subject to the same problems as the military eradication proposal.

In the end, after an intense week of discussions, the idea settled upon was to bring in advanced-warfare helicopters and trained sharpshooters who would take down the goats from the air. Most of the goats would be easily located by sight, but the rest would be tracked down with the use of Judas goats equipped with GPS monitors so that when these unwitting traitors found hidden goat communities, the helicopters would be close behind. Many of the Judas goats were females who had

---

[6] See http://www.darwinfoundation.org/english/_upload/isabela_atlas.pdf.

been given hormones to send off mating signals to unsuspecting males. This high-tech project was funded through an international collaboration that brought together the two island institutions—the GNP and the CDF—with USAID, the World Bank, the Global Environment Facility, Zanders Sporting Goods (for the automatic weapons), and several smaller donors. A total of 140,000 goats were killed over 2 years with approximately US$18 million in funding. Although perhaps an extreme example, the Isabela Project illustrates the attempt to separate life-worlds—humans from the environment and invasives from the pristine realm of the "untouched" landscape (untouched except for the marauding goats that were exterminated for being in the wrong place at the wrong time).

## Control of Invasive Plant Species

Blackberry (*Rubus niveus*) and guava (*Psidium guajava*) are considered two of the most problematic introduced plant species in Galapagos due to their aggressive reproductive strategies. Gardener et al. (2010a, b) suggest that, in the rural areas of Galapagos, complete eradication of these species might not be possible due to excessive cost and limited access to private lands. Instead, "indefinite" control and containment of the extent and location of invasion might be the most viable solution. Since 2008, the Galapagos National Park Service has worked with farmers to provide tools, herbicides, training, and educational programs that will allow them to recognize and treat invasive species on their land. This "invasive-maintenance" project falls under the new GNPS Plan for Total Control, which places high priority on the transition area between the park limits and the inhabited areas.[7] As of the summer of 2009, eight households were participating in this project, but many more were expected to sign up in the coming months. The project is a three-way collaboration: the GNPS provides the training, tools, and chemicals (approximately US$70,000 as of July 2009), the municipality provides money for refreshments, and the farmers provide their labor. According to the head of the Resources Division at the park, the GNPS now works most aggressively around the urban areas and in the zones of "impact reduction" surrounding the agricultural and livestock areas in the highlands. The GNPS also now considers it a priority to support land use practices in the agricultural areas that might help to control the spread of invasive species. According to the GNPS document outlining its new approach, there is now a consensus among Galapagos institutional actors that agricultural policies need to be designed that take into account the "multifunctional and multidimensional role of agriculture" and promote sustainable rural livelihoods (SIPAE 2006: 4–6). As such, the GNPS now supports the following policies: the transfer of technologies applicable to the ecological conditions of the islands, fostering ecologically sensitive

---

[7] See the somewhat dated project description at http://www.Galapagospark.org/programas/desarrollo_sustentable_agropecuario_especies_invasoras.html.

production for both subsistence and profit, strengthening the institutional and associative structures within the agricultural sector, and controlling and eradicating the species and pests that affect agriculture, while helping farmers to manage and restore key soil, water, and energy systems.

In many ways, this project situates the conservation work of the GNP in new areas; the park is acting outside of its direct spatial jurisdiction (the 97% of the terrestrial area of the archipelago that is protected) to shape practices in the private properties under the governance of local municipalities. And yet, the change is not as drastic as it appears; the park is not conceptually reworking the border between nature and society as much as it is physically and symbolically moving that border forward by several hundred meters to include the farmland in the protected areas of the reserve.

## Ecological Footprints

While the previous two examples focus on target species, other conservation approaches focus on human activity. A focus on self-regulation is fundamental to this approach. Of all the institutions on the islands, the CDRS is probably most emblematic of a "fortress conservation" approach that separates humans and the environment. When pushed by the most recent crisis to reevaluate its approach to the social system on the islands, the station began the "Human Footprint" program, which is currently one of the station's three flagship programs.[8] This is a new program designed by Christophe Grenier, the station's first social scientist. Grenier intended to continue the station's tradition of conducting robust, mechanistic science, but instead of studying ecological processes in isolation, he would work to quantify a series of indices for social processes to help different groups on the islands (e.g., taxi drivers, farmers, tourism operators, and restaurant owners), assess, and then self-regulate their environmental footprint.[9] Reflecting a global push toward sustainability, which recognizes the presence and needs of inhabited environments and attempts to balance these with conservation imperatives (Chambers et al. 2000), the Human Footprints project is one of the station's new areas of concern (Mark Gardner, Director of Terrestrial Science, July 25, 2009, personal communication). The station increasingly recognizes the need to incorporate the social system into its analyses, but it is clearly difficult to change gears in practice (Gardener and Grenier 2011).

In November 2008, the organization held a workshop in Galapagos with participants from the various conservation organizations in Galapagos as well as invited international experts in the field of restoration ecology. Over the course of several days, the experts debated projects and programs to foster conservation in and of the highland areas.

---

[8] http://www.darwinfoundation.org/english/pages/interna.php?txtCodiInfo=85

[9] As of 2010, Grenier is no longer with the Charles Darwin Station.

The final report presented 13 projects, representing the key areas of research for the humid and very humid zones of the inhabited islands. The proposed project areas were grouped under three independent research themes: (1) the spatial distribution, function, and value of different vegetation states; (2) the process of degradation; and (3) a toolbox for restoration. The third theme was the most tightly linked to human activities although the station largely reserved its analysis for abandoned farm lands, arguing that these areas were the primary conduits of invasive species from the inhabited areas to the protected ones. The station's focus on changing human behavior represents a significant shift for the organization; the incorporation of social science reflects a new concern with the ways in which humans connect with the natural world. It is this inseparability that appears as both a potential weakness and strength; if people are intimately embedded in the natural world, they must choose to either destroy it or save it. For the station and for much conservation policy, recognizing the role of humans in protecting the environment means refashioning human subjects to become better stewards.

## Rural Environmentality

Agrawal (2005) introduced the term environmentality to describe the institutional and cultural technologies through which individuals develop an environmental consciousness aligned with nature protection, self-regulation, and collective resource governance. Such an approach is currently in place in Galapagos, through new projects that are attempting to bring farmers into closer collaboration with the Galapagos National Park and local grassroots organizations in an effort to align the concerns of agriculture and conservation. On the main inhabited island of Santa Cruz, there are approximately 1,200 farmers who own land in the highland agricultural zone. These farmers are incorporated into three primary towns: Santa Rosa, Bellavista, and Cascajo. There are also several unincorporated communities governed by *juntas* (committees) that sit on the periphery of the agricultural area. The highlands of Santa Cruz are classified as a humid zone (mean annual precipitation of approximately 1,845 mm) with soils up to 1 m deep of basaltic origin, well weathered, and sandy loam in texture (Wilkinson et al. 2005). The native vegetation in the highlands has been cleared for agriculture and grazing. With respect to farming conditions, the highlands receive water during the wet season, but groundwater is scarce and limits the crops that can be grown. Many farmers subsist on extensive cattle ranching although manioc, corn, watermelon, and tree fruits are also common. While most of the farmers have been there for only one generation, some are descendants of the original colonists in the early 1900s (and in the 1800s, although few of the families from that period remain). Land in the agricultural areas of Santa Cruz is privately held, a result of the waves of state-sponsored colonization that took place during the 1960s where the Ecuadorian state allocated 100 ha plots to people willing to come and settle this "national frontier."

Examples of new initiatives with farmers include the agreements with municipalities to reforest a native tree, scalesia (*Scalesia pedunculata*), intercropped with coffee. A major proponent of these projects is FUNDAR. Created in 2001, FUNDAR is a local Ecuadorian nonprofit organization with a permanent staff of five people. The organization "plans and executes projects for the creation of a new paradigm that integrates conservation and responsible development. We open spaces for discussion, debate and reflection for change. We promote personal development, equality, social and environmental ethics, participation and strengthening of local abilities."[10] FUNDAR is funded primarily by international conservation NGOs such as the World Wildlife Fund and the Nature Conservancy, but as mentioned earlier, the organization received a large grant from the European Union in 2008 to work on sustainable agriculture projects with local farmers. This project centers on a community garden within a nature preserve (called Pájaro Brujo) located in the agricultural highlands and maintained by FUNDAR in which agroecological production methods are used to grow vegetables and trainings in these methods are provided to local farmers.

By the summer of 2011, several families had participated in the garden project. FUNDAR activists were also hoping to create a local farmers' market for organic produce to instill local pride in fresh, local produce and to make agriculture a viable activity for both conservation and development. Without the option of making a profit from farming, there is no future for the sector. The majority of small farms are run by elderly people subsisting off of retirement funds and farming because it is a way of life, rather than a way of making a living. As one of the directors of FUNDAR said in 2011:

> The agricultural zone is key because of the need for capacity building and the role [the farmland] plays in invasive species. Farmers don't have training, they prefer to sell lands rather than work them… Owners of some of the small farms came here in the 20s, 30s, and 40s… They were the first colonizers, brought here by the government for territorial presence and national security. They got received 100 hectares each. But these were poor people, with little fluid capital, and they lacked the sufficient economic resources to manage so much land. They did what they could, and the rest of the area became prone to invasive species. People saw the benefit of selling part of the land, subdividing it to sell to foreigners for the "vacation farms." Agriculturalists are land rich but money poor. The agricultural zone will be lost over time.

Park officials argued that it was difficult to get permission to work in the farm areas, and it is only recently that collaboration has become easier. More money from local economies and administrative resources is being put into farmland conservation projects, usually done through agreements with local municipal leaders or associations. The agreements are what allow the work to go forward: "Otherwise, we wouldn't be able to do anything." These projects present the farmland of the islands as multifunctional, working landscapes wherein nature and society are mutually constituted in everyday practices of digging, planting, raising, and exchanging. It is not clear what kind of articulations these everyday practices are generating, but they represent recognition of the inability to separate

---

[10] From FUNDAR's website http://www.fundarGalapagos.org/portalj/index.php?option=com_con tent&task=view&id=12&Itemid=26.

nature and humans, and they give value to the various ways in which livelihoods and biodiversity conservation might emerge together if each were seen as crucial to the other.

## Leaving the Conclusion Open: Adaptive Radiation of Conservation Practice?

Conservation policy is often written out in boardrooms, discussed during workshops, or illuminated in laboratories, but it is enacted on the ground and embodied in local people and communities. In Galapagos, new policies are depicted, as in the article by Vince, as either "embracing" invasive species or continuing the fight to eradicate them. This language positions "dirty" or contaminated landscapes against pristine ones and ascribes a scientific value to each (Geist et al. 2011; cf. Raffles 2011). The reality is rarely as neat and effective as the policies on paper.

These brief examples of different conservation policies in Galapagos highlight diverse articulations of knowledge, ecology, and governance. Debates over the "best way" to do conservation on the islands are misleading; in fact, actors from conservationists and park rangers to local tourism operators and farmers are negotiating constantly from different material and social positions to shape policies for particular spaces or resources or people within what are too broadly thought of as "the islands." The multiplicity of these positions has over time given rise to both conflict and temporary resolution manifested in new policies or programs or conservation practices, what we refer to as "pulp fictions" of the appropriate relationship between people and nature. In the case of Project Isabela and the Plan for Total Control, attempts to separate people and nature manifest spatially in borders and high-modern technologies of containment and control. In the case of the Ecological Footprints, attempts to recognize the role of humans while privileging nature manifest in new subjectivities fashioned through the internalization of moral imperatives. In the case of the new agricultural projects promoted by FUNDAR and the GNPS, attempts to integrate humans and nature manifest spatially in diversified working landscapes and living borders. All of these positionings represent ongoing negotiations and temporary resolutions in a space characterized by discourses of crisis. They illustrate the difficulty of characterizing "the" approach to conservation in the islands and suggest that perhaps the archipelago is as much a living laboratory for social science as for the biological sciences.

**Acknowledgements** The authors have been working together in the Galapagos Islands since 2007 and would like to thank Steve Walsh and Carlos Mena, codirectors of the Carolina Institute for the Study of the Galapagos, for their dedication to the islands and support of the authors' research. The authors would also like to thank Patricia Polo and Elizabeth Hennessy for research support.

# References

Agrawal A (2005) Environmentality: technologies of government and the making of subjects. Duke University Press, Durham

Agrawal A and Sawyer S (2001) Environmental Orientalisms. Cultural Critique 45:71–108

Atkinson R, Rentería J, Simbaña W (2008) The consequences of herbivore eradication on Santiago: are we in time to prevent ecosystem degradation again? Charles Darwin Foundation, Servicio Parque Nacional Galapagos

Bassett TJ, Zuéli KB (2000) Environmental discourses and the Ivorian Savanna. Ann Assoc Am Geogr 90(1):67–95

Blaikie P, Cannon T, Davis I, Wisner B (1994) At risk: natural hazards, people's vulnerability, and disasters. Routledge, New York

Boersma PD, Vargas H, Merlen G (2005) Living laboratory in peril. Science 308:925

Braun B (1997) Buried epistemologies: the politics of nature in (post)colonial British Columbia. Annals of the Association of American Geographers 87(1):3–32

Bryant R (2001) Politicized moral geographies. Debating biodiversity conservation and ancestral domain. Political Geography 19:673–705

Chambers N, Simmons C, Wackernagel M (2000) Sharing nature's interest: ecological footprints as an indicator of sustainability. Earthscan, London

Cronon W (1995) The trouble with wilderness; or, getting back to the wrong nature. In: Cronon W (ed) Uncommon ground: rethinking the human place in nature. W. W. Norton and Co, New York, pp 69–90

Crush J (ed) (1995) The power of development. Routledge, New York

Escobar A (1995) Encountering development: the making and unmaking of the Third World. Princeton University Press, Princeton, NJ

Gardener M, Grenier C (2011) Linking livelihoods and conservation: challenges facing the Galapagos Islands. In: Baldacchino B, Niles D (eds) Island futures and development across the Asia-Pacific region, Global Environmental Studies, Springer, pp 73–85

Gardener M, Atkinson R, Rentería JL (2010a) Eradications and people: lessons from the plant eradication program in Galapagos. Restor Ecol 18(1):20–29

Gardener MR, Cordell S, Anderson M, Tunnicliffe RD (2010b) Evaluating the long-term project to eradicate the rangeland weed *Martynia annua* L.: linking community with conservation. Rangel J 32:407–417

Geist D, Harmann O, Snell H, Whelan P (2011) Galapagos Goal: Eradication of Invasive Species. Reply to Vince (2011) "Embracing Species," http://www.sciencemag.org/content/331/6023/1383/reply. Accessed Oct 19, 2012

Gonzalez JA, Montes C, Rodriguez J, Tapia W (2008) Rethinking the Galapagos Islands as a Complex Socio-Ecoloigcal System: Implications for Conservation and Management. Ecology and Society 13(2):13

Guyer J, Peters P (1987) Introduction to conceptualising the household: issues of theory and policy in Africa. Dev Change 18(2):197–214

Hardin G (1968) The tragedy of the commons. Science 162:1243–1248

Holt FL (2005) The catch-22 of conservation: indigenous peoples, biologists and culture change. Hum Ecol 33(2):199–215

Jokisch B, Pribilsky J (2002) The panic to leave: economic crisis and the 'new emigration' from Ecuador. Int Migr 40(4):75–101

Leopold A (1949) A Sand County almanac, and sketches here and there. Oxford University Press, Oxford

Lu, F, Valdivia G. Wolford W. Social Dimensions of 'Nature at Risk' in the Galápagos Islands, Ecuador, *Conservation and Society*

Macdonald T Jr (1997) Conflict in the Galapagos Islands: analysis and recommendations for management. Report for the Charles Darwin Foundation. Harvard University, Center for

International Affairs, Program on Nonviolent Sanctions and Cultural Survival, Cambridge, MA, USA, p 15

Myers JH, Savoie A, Van Randen E (1998) Eradication and pest management. Annu Rev Entomol 43:471–491

Neumann RP (2004) Moral and discursive geographies in the war for biodiversity in Africa. Polit Geogr 23(7):813–837

Ospina P (2006). Galapagos, naturaleza y sociedad: actores socialaes y conflictors ambientales en las islas Galapagos. Bibl Cien Soc 55. Universidad Andina Simon Bolivar, Quito, Ecuador

Ospina P and Falconí C (eds) (2007) Galápagos: Migraciones, Economía, Cultura, Conflictos y Acuerdos Universidad Andina Simon Bolivar: Quito

Ostrom E, Nagendra H (2006) Insights on linking forests, trees, and people from the air, on the ground, and in the laboratory. Proc Natl Acad Sci 103(51):19224–19231

Peet R and Watts M (eds) (1996) Liberation ecologies. Environment, development, social movements. Routledge, London

Peet R, Robbins P, Watts M (2011) Global Political Ecology. Routledge, New York

Pelling M (2003) Natural disasters and development in a globalizing world. Routledge, New York

Quiroga D (2009) Crafting nature: The Galapagos and the making and unmaking of a "natural library." Journal of Political Ecology: Case Studies in History and Society 16:123–140

Raffles H (2011) Mother Nature's Melting Point, http://www.nytimes.com/2011/04/03/opinion/03Raffles.html (accessed October 19, 2012)

Simberloff D, Genovesi P, Pysek P, Campbell K (2011) Science 332(6028):419

Sodikoff GM (2012) The anthropology of extinction: essays on culture and species death. Indiana University Press, Bloomington

Thompson EP (1971) The Moral Economy of the English Crowd in the Eighteenth Century. Past and Present 50(1):76–136

Turner M (1999) No space for participation: pastoralist narratives and the etiology of park-herder conflict in Southeastern Niger. Land Degrad Dev 10:345–363

Vince G (2011) Embracing invasives. Science 331(6023):1383–1384

Watkins G, Cruz F (2007) Galapagos at risk: a socioeconomic analysis of the situation in the Archipelago. Charles Darwin Foundation, Puerto Ayora, Province of Galapagos, Ecuador. http://216.197.125.185/pdf/GalapagosAtRisk2007.pdf

West P (2006) Conservation is our government now: the politics of ecology in Papua New Guinea. Duke University Press, Durham, NC

Wilkinson S, Naeth M, Schmiegelow F (2005) Tropical forest restoration within Galapagos National Park: application of a state-transition model. Ecology and Society 10(1): 28. [online] URL: http://www.ecologyandsociety.org/vol10/iss1/art28/

# Chapter 6
# The Double Bind of Tourism in Galapagos Society

Laura Brewington

*No existe desarrollo sostenible [There's no such thing as sustainable development]*

(Isabela Island hotel owner, 2008).

*Poaching remains a serious threat and eco-tourism an even more serious threat. The Galapagos are being destroyed by both poachers and eco-tourism*

(Sea Shepherds Captain Paul Watson, 2011).

## Introduction

"This is not what it means to be *galapagueño*."[1] As the park guard made his sad proclamation, he stood in front of the desiccated carcasses of three giant tortoises whose flesh had been scraped away from their torsos and feet. It was June 2009, and a team from the Galapagos National Park Service (GNPS) was traveling on foot to one of Isabela's isolated beaches and a protected area, located several kilometers from the town of Puerto Villamil. The tortoises had been dead for over three months. They had been placed in tree branches at eye level. The faded numbers painted on their shells indicated that they were born and raised at the island's breeding center. As they photographed the remains, the guards agreed that this was likely the work of members of an old Galapagos family who were thought to be responsible for 16 such deaths the previous year.

---

[1] In this chapter, the term "*galapagueño*" refers to a permanent resident of the islands but, in colloquial use, is often reserved for descendants of the original colonists. In general, "resident" will be used to distinguish legal permanent residents from migrants and visitors.

L. Brewington (✉)
Department of Geography, University of North Carolina, Chapel Hill, NC, USA
e-mail: laura@sdf.lonestar.org

S.J. Walsh and C.F. Mena (eds.), *Science and Conservation in the Galapagos Islands:* 105
*Frameworks & Perspectives*, Social and Ecological Interactions in the Galapagos Islands 1,
DOI 10.1007/978-1-4614-5794-7_6, © Springer Science+Business Media, LLC 2013

Acts such as this are less common than they were a decade ago, when high-profile conflicts between the fishing sector and policymakers erupted into violent demonstrations. Illegal activity is prevalent today though, not only on Isabela Island but archipelago-wide. Such behavior is often driven by resistance to measures that limit local development of the fishing, and now, more commonly, tourism sectors. Tourism, as the driving force of today's Galapagos economy, has become what Environmental Minister Marcela Aguiñaga called "one of the main threats to the health and integrity of Galapagos," in her opening speech at the Sustainable Galapagos Tourism Summit held in 2010. Although it is often called ecotourism, there are conflicting notions about how tourism in unique and fragile environments should be realized, which have brought the industry under recent scrutiny. Accelerating introductions of new species, migration and illegal activity have come in the wake of the tourism boom, questioning how Galapagos "ecotourism" really is.

Ecotourism should, according to Martha Honey (2008), be environmentally sound and small scale, providing equal benefits to conservation as it empowers and enriches the lives of local residents, but the sudden growth and expansion of the industry in Galapagos has transformed this economic activity into a threat to conservation and social practices. Uneven shares of tourism-generated wealth perpetuate old tensions between those who benefit from environmental regulations and those who do not. Galapagos society, therefore, is caught in a double bind: (1) to pursue economic success and (2) to do so in an environmentally responsible and legal manner. Across the archipelago, people are struggling to come to terms with these two, often contradictory, demands that privilege some and marginalize others in the shadow of the tourism boom.

This chapter examines the tourism industry in Galapagos critically, from its inception in the 1960s, dominated by live-aboard cruises, to the present day as island-based touring has gained momentum over the "floating hotel" model originally promoted by conservationists. While the economic implications of this shift have been described (Taylor et al. 1999; Taylor et al. 2003, 2006; Epler 2007), as well as the direct and indirect environmental impacts (de Groot 1983; Honey and Littlejohn 1994; MacFarland 1998; Cléder and Grenier 2010; Ouvard and Grenier 2010), the industry's social and cultural drivers have rarely been considered.

As the Galapagos tourism industry is one of the fastest growing economies in the world (Taylor et al. 2006), pertinent questions can be raised about its impacts on island society, including: (1) How do residents perceive tourism-related development, and to what extent are they participating in and benefiting from it? (2) Who controls and benefits from tourism facilities and infrastructure? Answers to these questions should clarify whether Galapagos "ecotourism" is contributing to responsible development by promoting economic success among local populations and ensuring environmental sustainability and social accountability.

This chapter addresses these questions through a blend of quantitative and qualitative inquiry based on research conducted in the islands between 2007 and 2011. Cluster analysis of a large resident survey ($n = 1,242$) conducted in 2009 identifies particular social and demographic characteristics among the resident population that are conducive to supporting particular types of development or conservation, and investigates illicit environmental behavior in the context of environmental restrictions and economic need. This chapter then examines formal and informal tourism activities being practiced within the islands and considers the benefits and costs of the current tourism model in the context of the long-term management and economic development of the archipelago.

## Development, Conflict, and Sustainability

Commercial Galapagos tourism in the form of "floating hotels" began in the 1960s when New York-based Lindblad Travel began offering multiday cruises on their 66-passenger ship, the *Lina A*. Quito companies Metropolitan Touring and Turismundial joined Lindblad to expand the market, and between 1974 and 1980, the cruise ship fleet grew from 13 to 42 (Honey 2008: 125). Land-based tourism began in the 1970s with the availability of interisland shuttles and small boats for charter (Epler 2007: 3), but by 1982, only 18 hotels archipelago-wide had a total capacity for 214 guests.

Throughout the 1980s, the demand for food and goods alongside population growth outpaced disjointed environmental regulations, whose implementation was stalled because there was no clear leadership entity. In spite of restrictive land and marine use zoning (97% of available land area is Galapagos National Park), no regulations have ever been put in place to control tourist numbers. The current tourism model is the result of rapid and uncontrolled terrestrial expansion that occurred largely in the last three decades. In the 1980s and 1990s, island entrepreneurs began offering more land-based options for budget travelers, including Ecuadorian citizens and backpackers, and the dollar-based tourism economy enticed farmers and fishermen to explore alternatives to their traditional livelihoods. By that time, 26 hotels could accommodate 880 guests, and 67 ships held over 1,000 berths between them (Epler 2007: 13, 16).

In the midst of this early tourism boom, conflicts were generated among the increasingly regulated fishing sector. Commercial fishing of sea cucumbers, in particular, divided the resident population as well as the Ecuadorian government. Against regulations were local *pepineros* (sea cucumber fishermen), fishermen from the Ecuadorian coast, and the Ministry of Industry and Fisheries, while scientists and the Ministry of Agriculture expressed their strong support. Attempts to control the fishery were, as Honey writes, "disastrous…On the morning of January 3, 1995, a group of *pepineros*, some masked and wielding machetes and clubs, blockaded the road to the national park headquarters and research station outside Puerto Ayora [on Santa Cruz Island]" (Honey 2008: 134). On other occasions, disgruntled fishermen set fire to thousands of acres of land and threatened to kill giant tortoises held as "hostages" (Honey and Littlejohn 1994; Snell 1996; MacDonald 1997).

Soon, tensions grew between local tour operators and agencies based on the mainland, which controlled the Galapagos tourism market. This assumed the nature of a battle between residents and "outsiders," as naturalist guide Mathias Espinosa recalls (personal communication 2008). The pushback from residents was met with resistance by the Ecuadorian government when the issuance of *cupos* (passenger/ berth quotas) for local tour operators was suspended at the same time that the local fishing sector was restricted, resulting in explosive riots and demonstrations. "If the government wanted to economically strangle the Galapagos population," said Christophe Grenier, former head of social science at the Charles Darwin Foundation (CDF), "it would not have done anything differently: all of the islands' productive sectors were smothered under the pretext of protecting the ecology" (1996: 421).

Troubling levels of violence led to the development of the 1998 Special Law for Galapagos, a complex set of articles designed to control population growth, elimi- nate commercial fishing inside the Galapagos Marine Reserve, and promote respon- sible tourism development. A significant portion of the law was created by Galapagos residents to protect their economic interests and cultural integrity. Following the law's passage, trade unions and civil society organizations became important sites within local industry for residents to influence political decision-making when, according to anthropologist Pablo Ospina, "it became necessary to oppose the hold that environmentalism had on the province" (2001: 21). Permanent Galapagos resi- dency was established, granting residents rights to employment and wages 75% higher than on the mainland. Incoming migrants are restricted to renewable, one- year temporary residency and 90-day visitors' visas, and residency is monitored via an electronic ID tracking system.

The Special Law was a landmark piece of legislation that, in part, sought to alle- viate residents' concerns about the security of their livelihoods with the influx of recent migrants. It also served to reframe concerns about the impacts of tourism on more general population effects. While institutions like UNESCO acknowledge tourism's tight linkage with human population pressure in the islands, many scien- tists and policymakers do not (UNESCO 2007).[2]

Perhaps owing to this fact, implementation of the Special Law with respect to the tourism industry has been weak. Tourism continues to bring about considerable change in the urban and rural landscapes of Galapagos, and little has been done to encourage responsible development. Economics, more than sustainability criteria, have dictated decision-making, resulting in a 9% annual increase in tourist visita- tion and 150% growth in the number of island hotels (Epler 2007), while only 45 individuals and corporations own the 83 luxury, standard, and day-tour vessels operating in the islands (Epler and Proaño 2008). At the same time, conservation measures in Galapagos have been uneven and restrictive to the local population. Research on conservation psychology and political ecology has shown how illicit environmental behavior can arise out of marginalization and resentment (Neumann 1998; Kaplan 2000; Robbins et al. 2006; Khan and Haque 2010), demonstrated by

---

[2] Representatives of Galapagos conservation organizations often consider park officials, tourists, and tourism acceptable human influences, but not local populations. Although they rightly point to a history of unsustainable resource extraction by local communities, they do not acknowledge the similarly poor environmental track record of tourism (cf. Terborgh and van Schaik 2002).

continued acts of resistance when the needs and desires of Galapagos residents conflict with conservation mandates.

Currently, around 170,000 annual visitors travel to the islands where over 20,000 people live (GNPS 2011; INEC 2010). "No one envisioned that the islands would emerge as one of the world's premier ecotourism destinations; that Galapagos tourism would contribute hundreds of millions of dollars to Ecuador's national economy, and in turn, that it would generate revenues and population growth in Galapagos exceeding anyone's wildest expectation," Epler concludes (2007: iii). The annual growth rate in the number of tourists between 2000 and 2006 was 14%, falling behind only Panama, El Salvador, and Guatemala in percent visitor increase in countries within the Americas (Proaño and Epler 2008). If that rate continues, in under a decade more than half a million people will visit the islands every year.

For many inhabitants of the archipelago, the lures of Galapagos tourism and economic prosperity are illusory. The cost of living in Galapagos is three times that of the mainland, and although imported supplies such as gasoline are subsidized by the government, other products are high priced and often limited in availability. Without potable water or wastewater treatment systems, residents frequently experience intestinal problems and skin diseases. Health-care facilities are not equipped to handle most medical needs beyond minor surgeries, but flying to the mainland for hospital attention is not a financial option for many. Because the majority of tourism-related income remains in the hands of wealthy mainland or foreign-based tour operators, per capita income in Galapagos increased by less than 2% per year between 1999 and 2005, due largely to migration-induced population growth. "In real terms," write Taylor et al. (2006), "income per capita almost certainly declined." In the meantime, the permanent resident population alone is projected to increase to over 100,000 by 2030, if current growth rates hold (Proaño and Epler 2008).

## Methods

Beyond reports produced by institutions operating in Galapagos, literature concerning modern Galapagos society places a heavy emphasis on the now-waning fishing sector (Honey and Littlejohn 1994; Andrade 1995; Moreno et al. 2000; Ospina 2005). Other scholarship focuses on the construction of a local *galapagueño* identity (Ospina 2001; Borja 2003; Ospina 2003; Ospina 2006) and migrant demographics (Bremner and Perez 2002a, b; Kerr et al. 2004). Studies of Galapagos tourism have been economic (Taylor et al. 2006; Epler 2007; Epler and Proaño 2008) rather than social. The overall goal of this chapter is to identify the social and cultural ways in which conservation and development measures intersect with resident interests.

Rather than forming a homogenous social group, Galapagos residents have diverse goals, ways of knowing the islands, and economic engagements. To examine what combinations of conservation/development attitudes arose most frequently among Galapagos permanent residents, a cluster analysis was performed based on existing household survey data collected in 2009 by the Ecuadorian Statistical

**Table 6.1** Summary of 15 variables selected for cluster analysis

| Survey measure | Responses | | |
| --- | --- | --- | --- |
| Collect trash at tourist sites | Yes | No | |
| Believe introduced species are a threat | Yes | No | |
| Number of tourists should grow | Yes | No | |
| Should live "*isleño*" lifestyle[a] | Yes | No | |
| Should conserve island nature long-term | Yes | No | |
| Quality of life in a World Heritage Site is: | Good | Average | Poor |
| Boat-based (cruise ship, multiday trips) tourism should: | Increase | Stay the same | Decrease |
| Land-based (hotel stays, day trips) tourism should: | Increase | Stay the same | Decrease |
| Fishing should: | Increase | Stay the same | Decrease |
| Land transport should: | Increase | Stay the same | Decrease |
| Mainland marine transport should: | Increase | Stay the same | Decrease |
| Island marine transport should: | Increase | Stay the same | Decrease |
| Island air transport should: | Increase | Stay the same | Decrease |
| Mainland air transport should: | Increase | Stay the same | Decrease |
| Construction should: | Increase | Stay the same | Decrease |

[a]An *isleño*, or island-based, lifestyle is promoted by conservation institutions and emphasizes low imports, less motorized transport, responsible development, etc.

Institute and the Galapagos Government Council. The aim of the clustering exercise was to develop a resident typology that characterized the diverse motivations, expectations, and circumstances surrounding development, encouraging or obstructing residents' engagement in island conservation.

The 2009 survey was conducted to obtain current measures of the quality of life, health, education, and economic well-being of the permanent resident population. Using proportional, single-stage random sampling, investigators selected 1,336 households from the 72 census sectors in the province, which included the populated islands of Santa Cruz, Baltra, San Cristobal, Floreana, and Isabela. Of those, 1,242 households were selected for this analysis based on completed forms for the head of the household. Archipelago-wide, the average age of the household head was 43, and males comprised 82% of the respondents.

The survey form asked respondents to indicate their opinions about particular indicators associated with beliefs about the environment and growth in the tourism industry. Fifteen variables were chosen as surrogates for attitudes about development and conservation (Table 6.1). A cluster analysis was performed on these nominal, anominal, and ordinal variables, and four clear typologies emerged from the data. The clustering algorithm analyzes means for each measure, grouping the data by minimizing the within-group response variance and maximizing between-group variance (Kaufman and Rousseeuw 2005). This facilitates group comparisons of the roles of other variables that were not included in the clustering algorithm, such as amenities and expenditures, quality of life, education, and migration information. Pair-wise testing for differences in mean values and frequencies for these interval and ordinal variables was conducted at the 0.05 significance level.

Cluster interpretation is based on cluster means, past and present trends of conservation and tourism development in Galapagos, and the economic and geographic contexts in which residents engage with the tourist industry. Explanation of the cluster groupings, along with information on current trends in tourism and development, is discussed through interviews conducted between 2009 and 2011 with residents, policy-makers, tourism operators, and representatives of conservation organizations.

# Results

## *Permanent Resident Typologies*

The clustering exercise revealed that overall, Galapagos residents agree with the need for conservation in the islands ($n=1,215$, 98%) and the preservation of an *isleño* life-style ($n=1,140$, 92%; Fig. 6.1). This represents a practical understanding of Galapagos as a source of residents' livelihoods and cultural legitimacy (Ospina 2006: 52). In this respect, many informants expressed a profound pride in their province while at the same time making clear their desire for greater mainland access and everyday comforts. This is reflected by the fact that three-quarters of respondents live in the coastal urban centers where they engage in the growing private and public sectors, rather than traditional activities such as farming and fishing.

The clustering algorithm condensed the 15 variables concerning attitudes about development and conservation into four clusters (Fig. 6.2). A development typology was assigned to each cluster based on group responses to questions included in the algorithm. *Expansionist*: The first cluster comprises over half ($n=673$) of the survey respondents included in this analysis and describes a strong motivation for development, through mainland and island transportation, tourism, and construction. *Isolationist*: The second cluster ($n=310$) is characterized by a desire for moderate tourism development, high construction, and a lower opinion of life in a World Heritage Site. *Moderate*: The third group ($n=102$) is the smallest cluster and expresses low to moderate interest in tourism and local development. *Conservationist*: The fourth group ($n=157$) seeks stabilization or decrease in most aspects of island growth.

Now that general typologies have been formed, the factors shaping permanent resident attitudes about conservation and island economic growth can be considered. Analysis of the clusters on variables not included in the clustering process provided interesting insights and facilitated further description of distinct resident types as identifiable categories (Table 6.2).

**Fig. 6.1** Images from a 2008 GNPS publication for children that emphasize the difference between *isleño* (*left*) and mainland-based (*right*) lifestyles

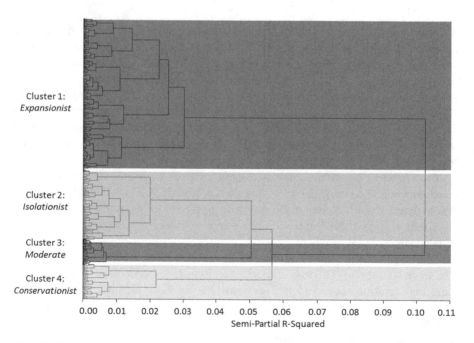

**Fig. 6.2** Dendrogram produced by the clustering algorithm. The four development typologies are indicated by alternating shades of *gray*

**Table 6.2** Survey information on household characteristics, education, amenities, and health

| Survey measure | Expansionist ($n=673$) | Isolationist ($n=310$) | Moderate ($n=102$) | Conservationist ($n=157$) | Signif[a] |
|---|---|---|---|---|---|
| *Household characteristics* | | | | | |
| Current residence | | | | | E,I–M,C |
| Santa Cruz/Baltra | 277 (41%) | 136 (44%) | 64 (63%) | 99 (63%) | |
| San Cristobal | 219 (33%) | 126 (41%) | 21 (21%) | 42 (27%) | |
| Isabela | 155 (23%) | 44 (14%) | 14 (14%) | 15 (10%) | |
| Floreana | 22 (3%) | 4 (1%) | 3 (3%) | 1 (1%) | |
| Household type | | | | | C–E,I,M |
| House | 442 (64%) | 197 (64%) | 62 (61%) | 120 (76%) | |
| Apartment | 87 (13%) | 38 (12%) | 13 (13%) | 23 (15%) | |
| Rented room | 67 (10%) | 38 (12%) | 17 (17%) | 9 (6%) | |
| Shack | 68 (10%) | 32 (10%) | 9 (9%) | 5 (3%) | |
| Other | 9 (1%) | 5 (2%) | 1 (1%) | | |
| Origin | | | | | |
| Galapagos | 178 (26%) | 74 (24%) | 16 (16%) | 36 (23%) | M–E,I |
| Sierra | 273 (41%) | 136 (44%) | 40 (39%) | 68 (43%) | |
| Coast | 215 (32%) | 96 (31%) | 43 (42%) | 51 (32%) | |
| Amazon | 7 (1%) | 4 (1%) | 2 (2%) | 2 (1%) | |
| Foreign country | 4 (1%) | 5 (2%) | 1 (1%) | 7 (4%) | |
| Years lived in Galapagos | 24.5 | 24.7 | 19.7 | 21.6 | M–E,I |
| *Education and employment* | | | | | |
| Highest education attained | | | | | C–E,I,M |
| None | 10 (1%) | 6 (2%) | 2 (2%) | 13 (8%) | |
| Primary | 34 (5%) | 10 (3%) | 2 (2%) | 4 (3%) | |
| Secondary | 244 (37%) | 118 (38%) | 41 (40%) | 29 (18%) | |
| Postsecondary | 281 (42%) | 122 (39%) | 43 (42%) | 50 (32%) | |
| College and above | 104 (15%) | 54 (18%) | 14 (14%) | 61 (39%) | |
| Job location | | | | | C–E |
| Local business | 309 (48%) | 126 (43%) | 51 (53%) | 86 (58%) | |
| Construction site | 47 (7%) | 30 (3%) | 11 (11%) | 7 (5%) | |
| Various sites | 121 (19%) | 41 (14%) | 11 (11%) | 22 (15%) | |
| Kiosk/street work | 7 (1%) | 5 (2%) | 2 (2%) | | |
| Local or rental property | 54 (8%) | 34 (12%) | 8 (8%) | 13 (8%) | |
| Domestic work | 25 (4%) | 22 (7%) | 4 (4%) | 9 (6%) | |
| Farm/ranch | 86 (13%) | 36 (12%) | 10(10%) | 12 (8%) | |
| *Spending and amenities* | | | | | |
| Monthly income needed to live well | $1,654 | $1,659 | $1,640 | $2,301 | C–E,I,M |
| Trouble paying for food during last 2 weeks | 189 (28%) | 62 (20%) | 31 (30%) | 28 (18%) | C–E, M |
| Household amenities | | | | | C–E,I,M |
| Many amenities | 489 (73%) | 235 (76%) | 74 (72%) | 132 (84%) | |
| Average amenities | 168 (25%) | 73 (23%) | 27 (28%) | 24 (16%) | |
| Few amenities | 16 (2%) | 2 (1%) | | | |
| Quality of life | | | | | C–E,M |
| Good | 92 (14%) | 62 (20%) | 18 (18%) | 44 (28%) | |
| Average | 531 (79%) | 228 (74%) | 78 (76%) | 106 (68%) | |

(continued)

**Table 6.2** (continued)

| Survey measure | Expansionist ($n=673$) | Isolationist ($n=310$) | Moderate ($n=102$) | Conservationist ($n=157$) | Signif[a] |
|---|---|---|---|---|---|
| Poor | 50 (7%) | 20 (6%) | 6 (6%) | 7 (4%) | |
| Current economic situation | | | | | C–E,I,M |
|   Able to save money | 79 (12%) | 43 (14%) | 13 (13%) | 36 (23%) | |
|   Equal save/spend | 344 (51%) | 177 (57%) | 62 (61%) | 84 (54%) | |
|   Forced to spend savings | 80 (12%) | 29 (9%) | 8 (8%) | 15 (10%) | |
|   Forced into debt | 170 (25%) | 61 (20%) | 19 (19%) | 22 (14%) | |
| Consider self poor | 306 (45%) | 122 (39%) | 43 (42%) | 34 (22%) | C–E,I,M |
| Play sports in last month | 302 (45%) | 140 (45%) | 50 (49%) | 91 (58%) | C–E,I,M |
| Internet access in last week | 121 (18%) | 59 (19%) | 27 (26%) | 74 (47%) | C–E,I,M |
| Amount spent on non-health mainland transport last 12 months | $171 | $153 | $188 | $270 | C–E,I,M |
| *Health* | | | | | |
| Sick last month | 286 (43%) | 136 (44%) | 44 (43%) | 88 (56%) | C–E,I,M |
| Has health insurance | 97 (14%) | 43 (14%) | 16 (16%) | 45 (29%) | C–E,I,M |
| Amount spent on health last 3 months | $108 | $122 | $131 | $195 | C–E,I,M |
| Amount spent on health last 12 months | $243 | $255 | $135 | $494 | C–M |

[a]Only variables with significant differences ($p<0.05$) in pair-wise testing are displayed (E for expansionist, I for isolationist, and so on)

*Expansionist*: The socioeconomic characteristics found in the first cluster are conducive for encouraging the most positive attitudes toward development. When cluster members were compared by residence, it was found that expansionists were the most highly dispersed across the urban and rural areas of the inhabited islands, with a higher concentration of "original" (Galapagos-born) residents than other groups. Pair-wise analysis of the frequency distribution was significant, suggesting that geographic distribution is associated with the respondents' attitudes about conservation. This is due in part to the strong representation of Isabela Island residents where, in spite of UNESCO recommendations, a new airport and dock were recently completed under the mantra, "*Isabela crece por ti*" [Isabela is growing for you]. This group is also characterized by the most ethnic diversity, the lowest overall quality of life, and is the most frequently forced into debt. Few (14%) have private health insurance policies, and little household income is spent on health-related issues.

*Isolationist*: Members of the second-largest cluster are concentrated on Santa Cruz and San Cristobal Islands, and the group is predominately located in urban areas. Like expansionists, they have a higher makeup of Galapagos-born residents than the other two clusters and exhibit the lowest attained education levels. They are characterized by a lower desire for tourism-related development than the expansionist cluster but express strong support for increased construction and transportation. This group has the lowest opinion of life in a World Heritage Site, and only 6% of respondents indicated that they collect trash at tourist sites. Households tend to have few amenities, and non-health-related spending is also the lowest in this cluster, but they experience greater job security than the other clusters.

*Moderate*: The third cluster is the smallest and contains the highest proportion of members originating from the mainland (85%), the majority of whom come from the coast. They migrated more recently than the first two clusters (average 19.7 years ago) and are more highly educated overall. However, they exhibit comparatively low awareness of the threats posed by introduced species, characteristic of those who migrated to Galapagos during the period of expansion in the 1990s (Heslinga 2003). The group is concentrated on Santa Cruz Island (63%) where they engage primarily in skilled labor and subsistence economic activities and experience the highest job security. Households have a moderate number of amenities, but report higher spending on health care and transportation to the mainland, and experience a low overall quality of life. They are characterized by a desire for some transportation improvements and boat-based tourism development, while most (68%) believe that land-based tourism should neither increase nor decrease.

*Conservationist*: The final cluster exhibits striking and statistically significant differences in development attitudes and socioeconomic characteristics from the other three. This group chiefly originates from Galapagos or the Sierra region of the mainland but has the largest constituent from foreign countries (4%). Many more are descended from foreign families and speak both Spanish and English. The cluster is predominantly urban and concentrated on Santa Cruz Island. High home ownership, very high education levels, low food insecurity, high savings and spending trends, and the most household amenities contribute to these respondents' experiencing the highest quality of life of any cluster. They are also the most likely to collect trash at tourist sites and express a strong desire for stable or decreased development, transportation, and boat-based tourism.

With the exception of the first cluster, survey respondents were in favor of stabilization or a decrease in the local fishing sector. Following the ban on industrial fishing in 1998, the sea cucumber and lobster fisheries virtually collapsed, leaving residents dependent on the less regulated, and less profitable, *pesca blanca* (whitefish) fishery. The coordination of fish sales to tour operators and sport fishing practices have been explored as alternatives to traditional fisheries that have met with limited success, particularly on Isabela Island. To this end, expansionists reported the highest participation in fisheries in the last 12 months of any cluster (8%), although this is still low compared to fisheries' activities a decade ago.

Given the disproportionately large share of the tourism economy that mainland tour operators hold (Taylor et al. 2006), it is not surprising that most residents are in favor of increased land-based tour development. Although Galapagos tourism is among one of the fastest growing economies in the world, only a fraction of total revenue (36%) remains in the islands (Taylor et al. 2006). The remainder is collected by large mainland touring companies who operate high-end cruises and own or rent passenger *cupos* (Epler 2007: 47). The Special Law granted permanent residents exclusive rights to obtain new tourism *cupos*, but this requires that they own a large boat that meets environmental

**Table 6.3** Distribution of visitors by to Galapagos by accommodation type in 2011 (Ecuadorian and foreign combined)

| Housing type | Number (%) of visitors |
|---|---|
| Hotel | 88,489 (48%) |
| Cruise ship | 78,447 (42%) |
| Family member | 13,199 (7%) |
| Private residence | 3,310 (2%) |
| Other | 1,583 (1%) |
| Total | 185,028 |

Source: (GNPS 2011)

regulations. Instead, locally owned pensions and hostels contract with fishermen and small boat owners for day tours (Honey 2008: 131). Even a third of conservationists, with significantly higher relative wealth than the other resident clusters, seek increases in land-based tourism. Indeed, island hopping is increasing in popularity over traditional "floating hotel" tourism: for the first time, in 2011, the number of visitors staying in hotels exceeded those staying on live-aboard cruise vessels (Table 6.3).

Much of the tourism-related infrastructure and development does not directly benefit residents, however. This reflects the fact that public services, particularly sanitary drinking and tap water, health care, and electricity, have been largely ignored during this period of growth.

Although Ecuadorian President Rafael Correa's administration has invested millions of dollars in mainland health care, marginal funding has been allocated to Galapagos. Limited access to sanitary water and sewer facilities frequently results in gastrointestinal and skin infections, especially among women, children, and the elderly (Walsh et al. 2010). None of the populated islands are prepared for serious viral outbreaks such as dengue fever (in 2005 and again in 2010) and H1N1 (2009), both of which arrived via tourists and visitors.

Growing problems such as crime and household waste are also attributed to the resident population. During the first five months of 2010, more than three-quarters (83%) of reported crimes in Puerto Ayora were committed by residents (Zapata, personal communication 2010). Santa Cruz Island, alone, generates 12 tons of waste per day, and although an estimated 35% of waste is recycled, the majority of is stored in a landfill until it is incinerated (Hardter, personal communication 2010). Despite the ubiquitous presence of trash canisters and recycling bins, littering persists in the islands' small towns. In a scathing editorial titled *The National Garbage*, American-born resident Jack Nelson writes, "This garbage doesn't come from offshore or Peru. It is not the kind of trash that falls from the hands of unthinking tourists. It is native, authentic island trash, lovingly Galapagos" (Nelson 2010: 4).

In light of increasing development and concerns about human impacts, the resident population has been the target of accusations that it is not capable of accepting the responsibility that comes with life in a World Heritage Site. Nelson has also attacked awareness campaigns by the GNPS, claiming that their portrayal of the *isleño* lifestyle is too abstract. Instead, he argues, residents must be told in no uncertain terms that what they are doing is environmentally unacceptable. Unfortunately, biodiversity goals rarely incorporate information from locals, and regulations are handed down as mandates. It is not uncommon to hear sentiments such as the following, expressed by one Santa Cruz resident, "They make us feel like we don't belong here—like the life of a giant tortoise is worth more than human life."

## Creating Sustainable Citizens

Unlawful environmental behaviors are acts of resistance by some residents, in response to restrictions perceived as external and illegitimate that have been imposed by conservation authorities. Such actions can be driven by need, while as Robbins et al. (2006) explain, "[S]ome is more overtly political." In part, authorities argue that increased surveillance and sanctions would stem unlawful activities, as the enforcement of environmental regulations in Galapagos has historically been minimal. The established penalty for engaging in illegal fishing includes a prison sentence ranging from 3 months to 3 years but is generally confined to confiscation of the vessel and a fine that is insufficient to deter future illegal activities. A seizure of $10,000 worth of shark fins may result in a fine of $2,000, a fraction of the value of one day's catch. Organizations like the Sea Shepherds, whose founder was quoted in the opening to this chapter, routinely push for greater application of sanctions within the marine reserve by the GNPS. A revision of the Ecuadorian Constitution in 2008 included a novel set of articles granting a unique set of rights to nature (Ecuadorian Constitution Article 71), which the Sea Shepherds urgently wish to apply to stop the poaching of endemic and native species that are protected by law (Emko, personal communication 2009). An exploration of illegal activity, however, necessitates an understanding of why residents would care for the environment in the first place.

To further capture reasons for environmental stewardship, a small opinion survey was conducted among 72 Santa Cruz and Isabela Island residents in 2010. Participants were asked to select one response out of four to the question, "Why would you participate in environmental protection?" and the results shown in Table 6.4 are paired with quotes from informants to further clarify the personal meaning of each statement. Those who responded, "It's unique in the world" or "Preserve it for future generations" adopt a view of the intrinsic value of Galapagos. They are represented by members of the conservationist cluster and are encouraged by conservation initiatives. As one young woman put it, "It's a privilege for us to live here, and it's our responsibility to protect it."

Members of the expansionist and isolationist clusters are more likely to agree with the majority (69%) of these respondents who chose a utilitarian view of the islands as a source of income or quality of life (responses 3 and 4). These clusters are comprised of more original families and the oldest migrants, a characteristic that Barber and Ospina (2008) also found to be related to a resistance to environmental

**Table 6.4** Residents' reasons for participating in conservation measures

| Survey response | Frequency | Quotes |
| --- | --- | --- |
| (1) It is unique in the world | 10 (14%) | "What we have in Galapagos, we don't have anywhere else" |
| (2) Preserve it for future generations | 12 (17%) | "In the future we want to see Galapagos like it has been, always" |
| (3) The environment is the source of our well-being | 26 (36%) | "*Galapagueños* have a very special identity. We care for our resources because we live from them" |
| (4) Good quality of life here | 24 (33%) | "Here I can still let my children go out to play without worrying" |

Source: Opinion survey, 2010

regulation. Their words express the pride in Galapagos that many residents share, intertwined with a sense of entitlement to the land.

It is that sense of entitlement, combined with hostility toward authority, however, that authorities fear is driving some residents to engage in unlawful environmental activities. In particular, there is an attitude among the "original" or "native" residents that they should not be subject to external regulations that are more concerned with plants and animals than people. For example, as one Isabela fisherman said in 2010, "the fish [populations] aren't a problem for us, for us the laws are the problem. To the conservationists everything we do is wrong." A marine comanagement scheme implemented through the 1998 Special Law was designed to facilitate the participation of fishermen in environmental decision-making, but its success has been tempered by a perceived lack of rights and access (Heylings and Bravo 2007) and punctuated by discoveries of illegal encampments along the coast (Suarez, personal communication 2010).

In contrast to clandestine fishing operations, highly visible infractions like the killing of giant tortoises are not fueled by a desire for or dependence on the use of protected resources. The reasons for resentment may include the rigid boundaries of the national park or the marine reserve, infringements on resource use rights, and perceptions of corruption among environmental managers or other environmental beneficiaries like tourism operators (Quiroga 2009).

As the cluster analysis reveals, Galapagos communities are not homogenous, and there are many reasons why residents would choose to support (or subvert, resist, and oppose) conservation regulations. The bitterness and disdain expressed by some informants stems from the awareness that funding destined for conservation projects will never benefit them. Measures that privilege the flora and fauna of protected areas over the needs and interests of their human counterparts generate further hostility among those poised to be conservation's greatest allies. In a final blow, the current model of development reinforces migrant flows from the mainland, a source of frustration for residents who argue that their interests were meant to be served by the 1998 Special Law.

**Table 6.5** Changing resident attitudes toward migrants between 2006 and 2010

| Survey response | 2010[a] | 2008[b] | 2006[c] |
|---|---|---|---|
| (1) Accept migration restrictions for family members | 42% | 47% | 43% |
| (2) Migrants result in environmental damage | 78% | 82% | 82% |
| (3) Migration increases local crime | 80% | 81% | 82% |
| (4) Migration increases local unemployment | 72% | 75% | 83% |
| (5) Migrants erode *galapagueño* culture | 89% | NA | NA |

[a]Source: Opinion survey (2010, $n=72$)
[b]Source: Barber and Ospina (2008, $n=302$)
[c]Source: Barber and Ospina (2007, $n=295$)

Contracted by hotels, high-end restaurants, and cruise vessels, skilled and unskilled migrants often fill employment needs that cannot be met by members of the resident population (Grenier 2007; Watkins and Cruz 2007). In this way, tourism supports the maintenance of a segmented labor force that requires migrants taking advantage of wage differentials between Galapagos and the mainland (cf. Massey 1999). This has also given rise to one of the few cases of domestic illegal migration in the world: an unknown number of these temporary migrants overstay their permits, thereby becoming illegal guests of the islands, of which there are an estimated 3,000 to 3,500 today (Sotomayor, personal communication 2010).

While social and environmental irresponsibility is frequently associated with the resident population in conservation discourse, residents see migrants as the source of the problem; perpetuating old inside/outside divides (Table 6.5). Residents tend to believe that unemployment due to the migrant influx is decreasing over time, but still express a strong agreement to the statement that migrants erode *galapagueño* culture, reflecting the sense of place described by each cluster above.

## *"Ecotourism": The Benefits and the Costs*

This chapter has highlighted the ways in which environmental management in Galapagos imposes legal restrictions on inhabitants, while perpetuating the conditions (and resident attitudes) that facilitate unregulated tourism growth. To quell accusations that mainland-based tourist agencies benefit from, but do not contribute to, the islands' welfare, some have begun to offer human services. Recognizing the difficulty and expense of medical transport to the mainland, for example, Celebrity Xpeditions instituted a program in 2010 to bring specialists to the Santa Cruz Island health center for week-long volunteer campaigns. Red Mangrove Galapagos and Ecuador Lodges, with hotels on three of the four populated islands, is developing family health and dental programs and assists with large-animal veterinary care on Isabela and Floreana islands.

Fundación Galapagos, an Ecuadorian for-profit organization founded by Metropolitan Touring, has promoted solutions in solid waste management for over 12 years.

Other organizations have attempted to address the fact that few local families are able to afford to explore the islands around them, meaning that the Galapagos archipelago's future leaders will scarcely know them. By 2009, Lindblad Expeditions and Metropolitan Touring had offered over 500 schoolchildren the opportunity to tour the islands on their cruises, a strategy that has boosted sales among foreigners, many of whom had no idea that up to half of the residents of Galapagos have never visited another island (Jenanyan, personal communication 2011).

Tourism has also provided an alternate source of income for residents who formerly engaged in illegal activity. Franklin, a former fisherman who came to Galapagos in the 1990s, guides day tours from Santa Cruz. But in the early years he lived on Isabela, participating in illegal shark fin, sea cucumber and lobster fisheries, and staging riots against the local GNPS office. "I was making $1,000 a day when my friends on the mainland were watching their money disappear. Of course I was going to keep doing it." Now he works in tourism, and he is happy with the change. "It's just not worth it. This is easier and I don't have to be looking over my shoulder" (personal communication 2010).

As mentioned above, to be an autonomous boat tour operator requires obtaining the right kind of boat and a *cupo*. Although the issuance of new *cupos* would promote community-based management and create a larger number of beneficiaries of tourism (Epler 2007: 48), a 2009 competition for the release of 72 *cupos* resulted in fewer than 20 proposals being approved (El Colono 2010a: 11). The process is particularly contentious on Isabela. While the current *cupo* system includes approximately 1,800 berths, they are exclusively owned by residents of Santa Cruz and San Cristobal islands. The presence of non-licensed tour operators also occasionally manifests in tragedy, as it did in early 2010 when two poorly equipped Isabela boats overturned while attempting to navigate the rocky entrance to a popular visitor site, resulting in serious passenger injuries on an island with only basic medical facilities.

The questionable legality of another tourism activity becoming popular among the islands' fishermen has generated recent conflict. Although *pesca deportiva*, or sport fishing, was prohibited by law in 2005 (Registro Oficial No. 564), operators claim that the GNPS and the Port Authority support sport fishing as a catch-and-release activity, a component of artisan fishing that is promoted as an alternative to commercial fishing. Proponents, including the mayor of San Cristobal, argue that it provides local fishermen with a tourist-based, sustainable alternative to traditional commercial fishing, with reduced pressure on local species. But, skeptics wonder, is this the kind of tourism that should be promoted in a place like Galapagos?

Although small operations by residents are expanding in the islands, the vast majority of tourism revenues and infrastructure remain in the hands of a few individuals and corporations (Epler and Proaño 2008). Large tourism operations have a seemingly limitless supply of lawyers and funding with which to defend their interests in the islands, while island-based operators, subject to the same conditions and requirements, are caught up in bureaucratic state control. The president of Metropolitan Touring, Roque Sevilla, is among the highest ranking executives in Ecuador but has been accused of diverting jobs from residents in the operation of his high-end Santa Cruz Island hotel, The Finch Bay, which employs primarily migrant workers (Zapata 2009: 2).[3]

The limited release of new *cupos* in 2009 further angered residents who see Quito-based operators like Metropolitan Touring with enough to support several yachts with over 100 passengers each (El Colono 2010b: 5). According to the mayor of San Cristobal, "Double talk doesn't work in Galapagos. ... It's obvious that [Mr. Sevilla] has his interests. He represents a group that has economic interests, that's who he is. I defend the public interest. ... Corruption can't be seen as something normal" (Zapata 2009: 2). The high-end "Iguana Crossing" hotel on Isabela Island generated similar opposition among residents when its mainland owner received permission from the Environmental Minister to build on top of a marine iguana nesting site. "This project was approved by the government," said Gardenia Flor, president of Isabela's Chamber of Tourism, "but it violates the desire of the community" (personal communication 2009).

Former GNPS director Raquel Molina refers to the network of large Galapagos tourism operators as the tourism "mafia." In March 2007, Molina was physically assaulted by members of the Ecuadorian Navy and Air Force as she and two park guards attempted to shut down an illegal kayaking operation on Baltra.[4] When asked about the conflict Molina responded, "They're corrupt, all of them. [Tourism operators] don't care about conservation in Galapagos—they care about making money... One day, eight major tour operators filed complaints about me at the municipality. I was just always in their way" (personal communication 2010).

The tourism industry itself has had its share of negative environmental impacts. As early as the 1970s, Silberglied noted that insects travel between populated islands and to distant sites on tour boats, a trend that has continued as pests and diseases are transferred with daily interisland ferry transport (Silberglied 1978). In 2001, an Ecuadorian tanker carrying diesel fuel, as well as bunker fuel that was destined for a luxury yacht owned by a mainland tour operator, ran aground off the coast of San Cristobal Island. Over 234,000 gallons of fuel were spilled into the waters that surround the archipelago's capital, Puerto Baquerizo Moreno, much of which was directed offshore by strong winds and currents (Fundación Natura and World Wildlife Fund 2001). In 2009, an Ecuadorian Navy training ship ran aground near Santa Cruz carrying 225,000 gallons

---

[3] Despite the fact that Finch Bay operates its own shuttle service and on-site farm, Sevilla recently argued that "licensed operators should be prohibited from vertical integration. In other words, tour operators should not be able to have their own on-land passenger transport service or be direct producers of food for tourists. This will allow more citizens to benefit from tourism as suppliers, even if they are not direct tourism service providers" (Sevilla 2008: 26).

[4] The altercation on Baltra was followed by Molina's 2008 dismissal from the GNPS by the Environmental Minister for insubordination, following her refusal to grant additional *cupos* to Sevilla.

of fuel, but was safely towed free (Arana, personal communication 2009). To date, however, cruise ships and day-tour boats do not undergo inspections or fumigations, and a contingency plan for environmental disasters like oil spills has never reached the draft stages (Rosero, personal communication 2011).

These issues raise critical questions about what kind of tourism model can best meet the islands' environmental and economic needs. Tourists also exert pressure on already-strained local resources, requiring food, water, and other commodities, in addition to the waste they generate. Many argue that this is a new kind of tourist, demanding amenities that can be found in the Caribbean or in Mexico: fine cuisine, discos, and luxury hotels. A writer for *Surfer Magazine* asked in 1998, "[O]n one of the great eco-tourism pilgrimages of all time, blessed with more intellectual raw data than perhaps anywhere on Earth: why are these clowns just doing the same bullshit they do at home?" (cf. Larson 2002: 234). That the naturalist guide pool has been increasingly "watered down" by new and lower-qualified guides is another indicator of the tour costs and quality that today's international tourists are seeking (Honey 2008: 157).

During the 2010 Sustainable Tourism Summit, workshop participants emphasized that the local culture is diverse and adapts to both internal and external forces, all clearly identified in Galapagos society, particularly as a result of the tourism boom of the past decade. As former CDF director Gabriel Lopez noted, "It's a major challenge to develop a shared vision for the common good among such a diverse community, but this is essential if we are to achieve a sustainable Galapagos." Proposals to double or triple the foreign entry fee to the national park (currently $100), initiate a lottery system, or limit visitors to one trip in a lifetime are some of the options proposed to control the exponential growth in visitor numbers, which UNESCO estimates will reach 400,000 per year by 2021 (Patel 2009). Paradoxically, as word spreads of the "crisis" in the islands, more people are compelled to visit them before it is too late (Neil 2008; Becker 2009; Bluestone 2009).

## Conclusions

Since the late 1980s, growth driven by the tourism industry has dramatically altered the social, political, and environmental realities of Galapagos. Given the changes tourism has brought to the archipelago over the last 30 years, can its trajectory of development be considered "ecotourism"? As Galapagos scholar Jane Heslinga cautioned in 2003, "Ecotourism, if properly monitored and managed, can contribute to environmental preservation through increased awareness, education, and financing. However, if inadequately regulated, ecotourism will degrade or destroy the ecosystems of globally significant areas" (Heslinga 2003).

Although on the surface the Galapagos Islands have been heralded as an international example of sustainable tourism (Honey 2008: 155), the goal of this chapter was to draw attention to the social and cultural aspects of a failed tourism model that has trapped Galapagos society in a double bind of development and sustainability. The 1998 Special

Law, intended to protect the interests of residents in light of new economic opportunities, has historically been weak in its implementation with respect to tourism. Migrant flows are reinforced by the industry, whose unrestricted growth places the increasingly restrictive measures on Ecuador's citizens in sharp relief. Physical control of terrestrial and marine visitor sites has left an estimated 95% of the archipelago's native flora and fauna intact, but exponential growth in the sheer numbers of people arriving every year threatens to undermine the national park's careful zoning. Finally, the indirect social and environmental effects of violations related to quarantine, permits, or safety threaten both inhabited and protected areas archipelago-wide.

A constituent of residents rejects and resists initiatives that they feel are imposed upon them and restrict their economic success. On the other hand, a small and affluent minority, aware of their dependence on tourism, has begun to "utilize the main symbols of science and conservation to further their particular cause" (Quiroga 2009). As such, it is critically important to recognize the trade-off between ensuring local benefits through development and ensuring that biodiversity goals are being met. According to former CDF director Graham Watkins, "Conservation can only work if the biodiversity in the archipelago is owned in the hearts and minds of those that live there. If the local community doesn't benefit economically from tourism, it's not going to support conservation" (personal communication 2008). The sustainability of Galapagos tourism remains very much in question, and the tenuous alliances formed among stakeholders have yet to assemble a coherent and egalitarian vision for the future.

# References

Andrade M (1995) Las comunidades pesqueras en la región insular. Charles Darwin Foundation, Puerto Ayora, Galapagos

Barber H, Ospina P (2007) Public acceptance of environmental restrictions. Galapagos Rep 2006–2007. 86–91

Barber H, Ospina P (2008) Public acceptance of environmental restrictions. Galapagos Rep 2007–2008. 40–45

Becker K (2009) Five more places to see before they are changed forever. http://www.gadling.com/2009/02/26/five-more-places-to-see-before-they-are-changed-forever. Accessed 27 Feb 2009.

Bluestone C (2009) See it before it disappears: reconciling and regulating disaster tourism. http://www.worldchanging.com/archives/010377.html. Accessed 27 Aug 2009.

Borja R (2003) Migraciones a Galápagos. Informe técnico de consultoría. Fundación Natura, Quito

Bremner J, Perez J (2002a) A case study of human migration and the sea cucumber crisis in the Galapagos Islands. AMBIO 31(4):306–310

Bremner J, Perez J (2002b) Demographic dynamics, gender, and resource use in the Galapagos Islands. InterCoast Winter. 20–35.

Cléder E, Grenier C (2010) Taxis in Santa Cruz: uncontrolled mobilization. Galapagos Rep 2009–2010:29–30

de Groot RS (1983) Tourism and conservation in the Galapagos Islands. Biol Conserv 26(4):300–300

Ecuadorian Constitution (2008). Article 71. Paragraph 1.

El Colono (2010a). Resumen 2009. El Colono Periódico de Galapagos. 11.

El Colono (2010b). Agua potable, cupos de turismo, y reforma educativa son las duedas del 2009. El Colono Periódico de Galápagos. 5.

Epler B (2007) Tourism, the economy, population growth, and conservation in Galapagos. Charles Darwin Foundation, Puerto Ayora, Galapagos

Epler B, Proaño ME (2008) How many tourists can Galapagos accommodate? Galapagos Rep 2006–2007:36–41

Fundación Natura, World Wildlife Fund (2001). The Jessica oil spill: God sleeps in Galapagos. In Fundación Natura, World Wildlife Fund, Galapagos Report 2000–2001, Quito.

Galapagos National Park Service (2011) Statistics of visitors to Galapagos. GNPS, Puerto Ayora, Galapagos

Grenier C (1996) Reseaux contre Nature. Conservation, tourisme et migrationes aux iles Galápagos (Equateur). Thèse de doctorat, l'Université de Paris Sorbonne.

Grenier C (2007) Conservación contra Natura. Abya Yala, Quito

Heslinga J (2003) Regulating ecotourism in Galapagos: a case study of domestic-international partnership. J Int Wildlife Law Policy 6:57–77

Heylings P, Bravo M (2007) Evaluating governance: a process for understanding how co-management is functioning, and why, in the Galapagos Marine Reserve. Ocean Coastal Manag 50(3–4):174–208

Honey M (2008) Ecotourism and sustainable development: who owns paradise? Island Press, Washington

Honey M, Littlejohn A (1994) Paying the price of ecotourism. Americas 46(6):40–48

INEC (2010) Censo de población y vivienda, Galapagos 2010. Instituto Nacional de Estadistica y Censos, Quito

Kaplan S (2000) Human nature and environmentally responsible behavior. J Soc Issues 56(3):491–508

Kaufman L, Rousseeuw PJ (2005) Finding groups in data: an introduction to cluster analysis, 2nd edn. Wiley, New York

Kerr S, Cardenas S, Hendy J (2004) Migration and the environment in the Galapagos: an analysis of economic and policy incentives driving migration, potential impacts from migration control, and potential policies to reduce migration pressure. Motu Economic and Public Policy Research, Wellington

Khan SMMH, Haque CE (2010) Wetland resource management in Bangladesh: implications for marginalization and vulnerability of local harvesters. Environ Hazards: Human Policy Dimensions 9(1):54–73

Larson EJ (2002) Evolution's workshop. Basic Books, New York

Macdonald T (1997) Conflict in the Galapagos Islands: analysis and recommendations for management. Harvard University, Cambridge

MacFarland C (1998) An analysis of nature tourism in the Galapagos Islands. Charles Darwin Foundation, Quito

Massey D (1999) International migration at the dawn of the twenty-first century: the role of the state. Popul Dev Rev 25(2):303–322

Moreno P, Murillo JC, Finchum R (2000) Diagnóstico socio—económico de las mujeres y familias del sector pesquero de Galápagos. Area de Educación y Comunicación Ambiental de la ECChD, Puerto Baquerizo Moreno, Galapagos

Neil D (2008) Please don't go. The Los Angeles Times. http://articles.latimes.com/2008/jun/01/magazine/tm-800words06. Accessed 17 June 2008.

Nelson J (2010) La basura patrimonial. El Colono Periódico de Galápagos 4.

Neumann RP (1998) Imposing wilderness: struggles over livelihood and nature preservation in Africa. University of California, Berkeley

Ospina P (2005) Las organizaciones de los pescadores en Galápagos. Instituto de Estudios Ecuatorianos, Quito

Ospina P (2001) Migraciones, actores e identidades en Galápagos. Consejo Latinoamericano de Ciencias Sociales, Quito

Ospina P (2003) El hada del agua: ética ambiental y actores sociales en Galápagos. J Intercult Stud 30:59–59
Ospina P (2006) Galápagos, naturaleza y sociedad. Corporación Editora Nacional, Quito
Patel T (2009) Immigration issues in the Galapagos Islands. Council on Hemispheric Affairs, Washington
Ouvard E, Grenier C (2010) Transporting passengers by lanchas in Galapagos. Galapagos Rep 2009–2010:40–47
Proaño ME, Epler B (2008) Tourism in Galapagos: a strong growth trend. Galapagos Rep 2006–2007:31–35
Quiroga D (2009) Crafting nature: the Galapagos and the making and unmaking of a "natural laboratory". J Polit Ecol 16:123–140
Robbins P, McSweeney K, Waite T, Rice J (2006) Even conservation rules are made to be broken: implications for biodiversity management. Environ Manag 37(2):162–169
Sevilla R (2008) An inconvenient truth and some uncomfortable decisions concerning tourism in Galapagos. Galapagos Res 65:26–29
Silberglied RE (1978) Inter-island transport of insects aboard ships in the Galapagos Islands. Biol Conserv 13:273–278
Snell HM (1996) Conservation gets personal. Noticias de Galápagos 56:13–16
Taylor JE, Dyer GA, Stewart M, Yunez-Naude A, Ardila S (2003) The economics of ecotourism: a Galapagos islands economy-wide perspective. Econ Dev Cult Change 51:977–997
Taylor JE, Hardner J, Stewart M (2006) Ecotourism and economic growth in the Galapagos: an island economy-wide analysis. University of California, Davis, Giannini Foundation of Agricultural Economics, Davis.
Taylor JE, Yunez-Naude A, Becerril J, Dyer Leal G, Martínez-Huerta ML, Ruiz M, Stewart M (1999) Estudio económico de Galápagos: Informe Inicial. Banco Interamericano de Desarrollo, Washington
UNESCO (2007) State of conservation of World Heritage properties inscribed on the List of World Heritage in Danger. United Nations Educational, Scientific and Cultural Organization, Paris
Walsh SJ, McCleary AL, Heumann BW, Brewington L, Raczkowski EJ (2010) Community expansion and infrastructure development: implications for human health and environmental quality in the Galapagos Islands of Ecuador. J Lat Am Geogr 9(3):137–159
Watkins G, Cruz F (2007) Galapagos at risk: a socioeconomic analysis of the situation in the archipelago. Charles Darwin Foundation, Puerto Ayora, Galapagos
Zapata P (2009) Alcalde de San Cristobal crítica doble discurso de Roque Sevilla. El Colono Periódico de Galápagos 2.

# Chapter 7
# The Evolution of Ecotourism: The Story of the Galapagos Islands and the Special Law of 1998

Michele M. Hoyman and Jamie R. McCall

*Unfortunately it [the Special Law of 1998] was ahead of its time. We the people of Galapagos were not ready to accept all the responsibilities the law gave us which is a pity. The law remains one of the best pieces of legislation for any protected area in the world and Ecuador should get credit for passing it because we had to change the Constitution to get it. Not a single other country in the world has done that. Eliecer Cruz*

(Bassett 2009, 84)

## Introduction

Every community has an economic development story to tell. These stories are as varied as the actors who make up the community itself—its citizens, business leaders, nonprofit organizations, etc. The interactions between these political and civic actors and their governments have a great deal of impact on economic development policy decisions and strategies. The above quote demonstrates the profound impact of the Special Law of 1998 on communities and citizens of the Galapagos. For many decades now, public and nonprofit sector leaders in the Galapagos Islands have greatly expanded their efforts to find a balance between economic development and environmental conservation. The crowning piece of legislation created from these

M.M. Hoyman, Ph.D. (✉)
Department of Political Science, University of North Carolina at Chapel Hill,
Chapel Hill, NC, USA
e-mail: Hoyman@unc.edu

J.R. McCall, M.P.A.
Department of Public Administration, North Carolina State University,
Raleigh, NC, USA
e-mail: jrmccall@ncsu.edu

S.J. Walsh and C.F. Mena (eds.), *Science and Conservation in the Galapagos Islands:*     127
*Frameworks & Perspectives*, Social and Ecological Interactions in the Galapagos Islands 1,
DOI 10.1007/978-1-4614-5794-7_7, © Springer Science+Business Media, LLC 2013

efforts is the Special Law of 1998, and the law is in many ways the foundation of the modern Galapagos economic development story.

The story of the Galapagos Islands is one of a tourism-based economy that has grown dramatically for nearly half a century. No aspect of development policy for the Galapagos Islands is more important than its approach to ecotourism, due to both its status as a habitat for numerous endangered animals and plants and its historical legacy due to the research of Charles Darwin. In many ways, the use of ecotourism as an economic development strategy represents the ideal solution to managing the environment and simultaneously bringing economic prosperity to the islands' residents. By ecotourism, we refer to the strategic use of environmental resources to promote environmentally conscious development and to ensure that tourism promotion is based on conserving the natural resources of the area (Blamey 2010; Bjork 2000; Donohoe and Needham 2006; Einarsdottir et al. 2012; Page and Dowling 2002; Valentine 1993; Fennel 2001, 2003).

As of 2012, many of the key ecotourism regulatory policies, such as the Special Law of 1998, are being scrutinized and updated. Ecuador is on the eve of redrafting a new law to replace the Special Law of 1998. Thus, now is a propitious time in the "evolution" of ecotourism in the Galapagos Islands to analyze the success and failures of government policies, using both a political and economic development framework. There are many international examples of economies based on ecotourism—including places like Belize, Thailand, and Nepal. Out of all these places, why choose the Galapagos to study ecotourism? Because in no other country has there been a government policy that is as ambitious or comprehensive as the ecotourism policies set by the Special Law of 1998. Enough time has elapsed— almost 15 years after its initial passage—to make an assessment of its impact in terms of social implications. The features of the Special Law are multifaceted; they cover everything from quarantines on invasive species to prohibiting mainland Ecuadorian in-migrants from holding jobs or having fishing licenses unless they first achieve permanent residency status. The scope of the Special Law's legislation is so sweeping that it represents a rare find in the public policy realm—few governments would go this far in the name of sustainability.

As with most public policies, development of the Special Law of 1998 was influenced by many cultural, political, and social forces. However, one of the most dramatic events which spurred the development of the bill was the near extinction in the late 1990s of many types of local wildlife, including sea lions, cormorants, and spiny lobsters. At the same time, the government was managing numerous cases of invasive species brought into the Galapagos by tourists or immigrants (Bassett 2009). One particularly notable example was the accidental introduction of feral goats into the environment that had a main diet of cactus plants, which they found themselves having to share with the native giant turtles called Tortugas. There were simply not enough of the plants left to be the main diet for two species, and as a result almost 200,000 goats had to be eliminated at great cost (Bassett 2009, 88).

Many lenses could be used to analyze the impact of the Special Law of 1998. In this chapter, we will use the lens of the opinions of local leaders. In July 2011, a number of local government and nonprofit leaders were interviewed in San Cristobal and Santa Cruz on their perspectives on the Special Law. We use the lens of leaders because their positions make them inherently familiar with ecotourism policy, and their perspectives give them a firsthand account of what is working and what needs improvement. Respondents to the survey discussed at length which aspects of the Special Law were effective and which were not. We will analyze their opinions on these matters later in the chapter.

This chapter proceeds in the following manner. The first section covers the historical, political, and economic context of the Galapagos including its unique status within Ecuador, its institutions, and the political conflicts among various actors seeking to shape public policy. The second section is a brief description of the legal environment of the area, including the main features behind the Special Law of 1998. The third section presents our findings on the degree of support local leaders have for the Special Law's provisions and the leaders' assessment of the effectiveness of those provisions. The fourth and final section presents a "model" ecotourism policy based on the interviews with local leaders.

## The Historical, Economic, and Policy-Making Context of the Galapagos Islands

The protection of the indigenous wildlife and flora of the Galapagos Islands has been an issue in the international spotlight for decades. The islands' status historically can be traced to Charles Darwin's famous expedition to the area and his subsequent evolutionary treatise, *On the Origin of Species*. The islands are an alluring destination for naturalists, photographers, and tourists who want to experience the pristine setting and unique animal life. Between the 1960s and 1990s, the number of visiting tourists exploded, as did the population of the islands.

Because of the Galapagos' place in the international dialogue on sustainability, numerous international nonprofit groups play a significant role in local affairs and politics. The presence of these groups—and the potential they have to shape ecotourism policy—is yet another facet of the islands that makes them a unique case study. There are numerous large and powerful international organizations involved in Galapagos policy making. For example, the United Nations Educational, Scientific and Cultural Organization (UNESCO) worked in conjunction with Ecuadorian President Correa[1] to develop key parts of the Special Law.

---

[1] A landmark event of note was Ecuadorian President Correa's proclamation on April 10, 2007, that the ecosystems of the Galapagos Islands were endangered. This was followed by UNESCO's decision to include the Galapagos on the list of World Heritage Sites that were endangered, with the familiar litany of threats to its ecology: too many points of access, too many immigrants, and not strict enough enforcement of quarantines. UNESCO's endangered designation was lifted in 2010, evoking mixed reactions. Although some viewed it as a tangible sign of progress, it also meant that it might be more difficult to raise money for conservation in Galapagos.

*Tourism remains the foundation of the Galapagos economy.* The economy of the Galapagos is somewhat diversified, including fishing and agriculture, but tourism remains the main economic driver. Quantitative analysis of the economy demonstrates this point. Scholars like Epler (2007, 27) and Taylor (2006, 140) have found that an increase in tourism activity by 10% leads to a concurrent increase in the agriculture industry by 3.9% and an increase of 4.7% in the fishing industry (Epler 2007, 27). Estimates of the tourist multiplier vary, but Epler (2007) estimates that for every $1 spent on tourism, $0.22 is injected into the local economy.[2]

*Galapagos residents do not have a culture of valuing conservation.* The residents of the Galapagos Islands, except for members of the political elite such as local government leaders and conservation foundation leaders, espouse a "use it now, worry later" mentality. This approach is the exact opposite of a culture of conservation and works from a presumption that the beauty and nature will be there forever. MacDonald (1997, 3) finds that the residents have a sense of entitlement to local resources, and they view government policies as "alien, imposed, and inappropriate." The feelings of marginality on the part of residents only fuel further resentment toward government policies.

*Regulating island in-migration is a challenge.* The Galapagos Islands' history and position in the international spotlight affords them an ever-increasing number of tourists, which increases the number of tourism jobs available. Concurrently, many parts of island life are heavily subsidized. The combination of these two factors presents an alluring opportunity for potential migrants—high-quality jobs with a relatively (subsidized) low cost of living. This combination inevitably leads to more immigration than Galapagos can realistically manage. For example, in 2007, it was estimated that for every 24,000 legal residents of Galapagos, there were 1,800 temporary residents and 5,500 illegal residents (Epler 2007, 36). The low level of government enforcement makes it difficult to deport illegal residents once they gain entry to the Galapagos (Epler 2007, 38). The migration issue is particularly salient because evidence shows that the wages of nonresidents do not generate as much local economic activity as do the wages of residents (Epler 2007, 38), so in theory, there is both an economic and political imperative to reduce immigration.

*The institutions of the Galapagos play an important role in shaping policy.* When it comes to crafting social and economic policy, scholars have long established that institutions matter. Disparate scholars have all agreed on the political and economic importance of institutions (Barzelay and Gallego 2006; North 1990; Knott and Miller 1987). Galapagos institutions share several common characteristics: they are fragmented, they are plural, they lack capacity, and the local institutions are overlaid by international organizations. Local government leaders in the Galapagos described all these characteristics to the authors. For example, one respondent said,

---

[2] This highlights one of the central criticisms of tourism-based economies—much of the revenue spent on tourism does not find its way back to the local community. In the case of Galapagos, approximately $0.88 of every dollar spent ends up going to outside business or overseas corporations.

"the institutional design is a problem," and another said, "enforcement is not strong enough." Another leader captured the consensus when he said there is "need for more regulatory power."

The creation and enforcement of ecotourism policy in the Galapagos takes place in a complex setting involving multiple layers of government organizations (Watkins and Cruz 2007). An important policy-making organization is the Galapagos National Park (GNP), created in 1959. The jurisdiction of the GNP encompasses 97% of the Galapagos land area and the Marine Reserve, which surrounds the shores of the islands. The Charles Darwin Foundation (CDF) was also created alongside the GNP, and is considered a "sister agency." The CDF provides research and advisory services to the park itself. The GNP represents one of a web of institutions that must cooperate in order to enforce ecotourism policy, but this network of institutions has undergone some notable changes as of 2012. The Galapagos National Institute (INGALA), an Ecuadorian agency that previously tabulated and controlled immigration into the Galapagos Islands, has been dismantled. The Governing Council of the Galapagos has taken over the power and responsibilities of INGALA. These government institutions coexist alongside powerful local elected officials such as the mayors of Santa Cruz and San Cristobal as well as provincial-level leaders. There are also several codetermination bodies in the Galapagos, which may require multiple rounds of consensus building before acting. One example of this is the Participant Management Board[3].

Among the important international nonprofits that play a role in shaping environmental policy are the World Wildlife Foundation (WWF) and UNESCO. Local nonprofits are also beginning to have an increasingly prominent role in shaping policy. One such example is Calidad Galapagos—a nonprofit that grew out of the islands' Chamber of Commerce—which inspires green business certification and gives out awards of differing star ratings based on businesses' use of energy-efficient light bulbs and other conservation measures. Another example is I.C.E., Immerse–Connect–Evolve, a group that works alongside the Santa Cruz municipal government to provide various community services, such as English as a Second Language classes and healthcare information. The university with the greatest permanent presence in the area is the Universidad San Francisco de Quito's (USFQ) Galapagos Academic Institute for the Arts and Sciences (GAIAS). Recently, the University of North Carolina at Chapel Hill's Galapagos Science Center opened on a campus adjacent to GAIAS, in collaboration with USFQ.

The final group of important institutional actors includes the tourism and fishing industries, which make up the bulk of the islands' economic productivity. Interests within the islands' business sectors are diverse and range from local fisherman running small businesses to corporate executives operating international cruise lines that visit the area. This makes for a highly complex mixture of actors and shared power especially when, as we will discuss later, many of the interest groups are advocating for different public policy outcomes.

---

[3] This organization was set up to allow all relevant parties to share in decision making and problem solving regarding the Galapagos Marine Reserve, which was established in 1989 and expanded several times afterward.

*The policy environment of the Galapagos is characterized by persistent conflict.* The previous section described three groups of key policy-shaping institutions and actors: government institutions and elected leaders, nonprofit and educational organizations, and business leaders. These groups represent powerful interests, which often result in competing policy goals. There are three main dyads of organized interests that are most likely to conflict: fishing interests versus conservation organizations, tourism interests versus conservation organizations, and international organizations versus local organizations. For decades, these groups have had ongoing disagreements on the balance between conservation and growth in Galapagos. These schisms among different interest groups form the context in which ecotourism policy decisions must be made and inevitably impact policy outcomes like the Special Law of 1998.

It is perhaps not surprising to find that the fishing industry and conservationists disagree on the balance between development and environmental preservation. At the local level, fishing industry representatives tend to view the Galapagos National Park as an adversary. The conflict between these organized interests often breaks out in violence. For example, the fishing industry is blamed for a wide-scale slaughter of Tortugas and for intentionally setting a fire on park land on the island of Isabela (Epler 2007, 39). According to the World Wildlife Fund, fishermen constitute only 3% of the population but are responsible for a large volume of economic activity (World Wildlife Fund, 2012). Fishing interests, though small in number, are highly mobile, distrustful of conservationist groups like the Charles Darwin Foundation, and militant in their behavior. Nonetheless, they are organized into multiple balkanized collectives with no central structure or identifiable leader (Quiroga et al. 2009, 117). In practical terms, this means it is difficult to negotiate with the fishing interests.[4]

A second persistent conflict among interest groups is the tourism industry versus conservation organizations. Tourism interests are particularly diverse because tourist organizations vary so widely in size and scope. At one end of the scale are innumerable "mom and pop" shops, often employing family members. These small-scale operations vary in their operations from island-hopping tours to selling T-shirts and other souvenirs. At the other end of the scale are a handful of multinational corporations that operate cruise liners in the area or have luxury hotels on the islands. Although conflict between tourism interests and conservation organizations is less dramatic and visible than some of the other schisms we have explored, it is nonetheless one of the most important schisms.

A final schism is between the local institutions and the international organizations. There is often a difference in values between local actors and international

---

[4] The driving force behind the conflict for the fishing industry is the international market. There is a thriving international market for sea cucumbers and shark fins, and the high prices buyers are willing to pay for these items are a clear temptation to break island conservation laws. It is simply irrational for the fishermen to cut back voluntarily on lucrative catches in these markets. However, the impact of overfishing is becoming clear. There are numerous studies documenting that supplies of overfished commodities are diminishing precipitously (Hearn et al. 2005).

**Fig. 7.1** Historical
distribution of Galapagos
National Park Fees, 1998.
Source: Galapagos National
Park, 2011

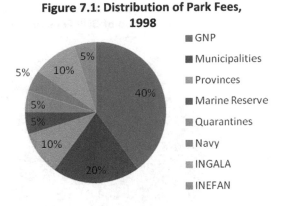

Figure 7.1: Distribution of Park Fees, 1998

**Fig. 7.2** Current distribution
of Galapagos National Park
Fees, 2011. Source:
Galapagos National Park,
2011

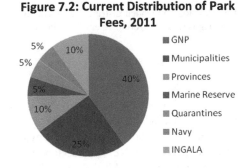

Figure 7.2: Current Distribution of Park Fees, 2011

groups in terms of their vision for the future of the islands and the balance between economic growth and sustainable conservation. Put simply, local residents have to live in the Galapagos on a day-to-day basis. Although they may value conservationist principles, they have more immediate needs that can only be achieved by economic development. In contrast, international groups often treat the Galapagos as if it were only a nature preserve and minimize the immediate needs of residents.

*Funding allocations to government institutions.* Many key government institutions receive at least partial funding from the fees collected by the Galapagos National Park. These fees are important because they represent a stream of revenue that is not affected by political decision makers and other factors. Figure 7.1 demonstrates fee distribution as of the 1998 Special Law and Fig. 7.2 shows the distribution as of 2011.

## Figure 7.3: Survey Response Recommended
## Distribution of GNP Fees to Different Institutions

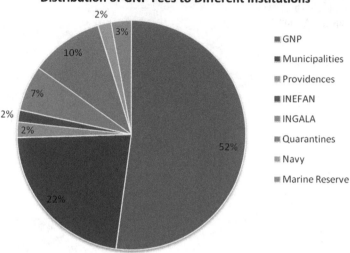

**Fig. 7.3** Recommended distribution of Galapagos National Park Fees by respondents. Source: Galapagos Local Government and Non-Profit Leaders Survey Research, 2011

As Fig. 7.1 demonstrates, in 1998, the park retained the plurality of fees. Local governments, provincial governments, and INGALA also received a share of the fees. Figure 7.2 shows that a few things changed by 2011. The Institute of Forests and Natural Areas (INEFAN) no longer exists, and other institutions have assumed its governing responsibilities, including the Ministry of Environment of Ecuador. The fee portion formerly received by INEFAN has been transferred to local municipal governments. In the course of interviewing Galapagos local government and nonprofit leaders about ecotourism policy and the Special Law of 1998, we asked them for their perspective on how fees should be distributed among government institutions. Figure 7.3 shows the results. As Fig. 7.3 demonstrates, local leaders have a notable level of disagreement with how park fees should be distributed.[5] On average, leaders believe that the park should receive the majority of fees. They indicated that the park should receive 52% of the total fees, which would be 12% above 2011 funding levels. Also, leaders would significantly decrease funding for

---

[5] It should be noted that Fig. 7.3 is an *average* of all respondents' views, and it does not give one the full sense of the huge range of views that leaders espoused. For instance, on one end of the continuum, one respondent stated, "The Park has too much power." On the other end of the continuum, another respondent said, "The Park has too much responsibility and not enough power."

**Table 7.1** Summary of changes in fee distribution

| Institution | 1998 fee | 2011 fee | Difference | Leader difference |
|---|---|---|---|---|
| National Park | 40 | 40 | 0 | −12 |
| Municipalities | 20 | 25 | +5 | −3 |
| Provincial | 10 | 10 | 0 | −8 |
| Marine Reserve | 5 | 5 | 0 | −5 |
| Quarantines | 5 | 5 | 0 | +5 |
| Navy | 5 | 5 | 0 | −3 |
| INGALA[a] | 10 | 10 | 0 | −3 |
| INEFAN | 5 | 0 | −5 | +2 |

[a]As of 2012, INGALA has been replaced by the Council of Galapagos Governments

provincial governments (decreasing funding from 10% to 2%). Table 7.1 above summarizes the data from Figs. 7.1, 7.2, and 7.3, noting the differences between the 1998 level, the 2011 level, and the level recommended by the leaders.

## The Legal Environment of the Galapagos Islands

The Galapagos Islands have long maintained special status with Ecuador's federal government as a favored province. This status has granted residents of the islands certain economic benefits, some of which were enhanced by the Special Law of 1998, that are not offered to other provinces (WWF 2003, 78). The main subsidies that were either brought into effect or enhanced with the passage of the 1998 Special Law include:

- A 50% discount on airfare for travel between the Galapagos and mainland Ecuador for residents (WWF 2003, 54)
- Two wage subsidies for Galapagos-based workers: (1) a super wage, 75% higher than mainland wages for all jobs, and (2) a teachers' minimum wage, which is an additional 75% higher than the Galapagos minimum wage (Special Law 1998)
- Subsidies for gasoline purchases on the islands (Jácome 2008)
- Subsidies for electricity and water on the islands (the costs of which are very high due to the technical challenges of delivering effective and environmentally safe utilities to residents) (Jácome 2008)

Due to the way in which these subsidies are designed, they continue to grow as the islands' population increases. If the appetite for ever-expanding growth in Galapagos is not stemmed, these subsidies could rise to a very high level. These public policies are not without their critics—some claim the use of subsidies distorts the market by driving up demand-related tourism jobs and by making immigration grow at a faster rate than it otherwise would (WWF 2003, 80).

Perhaps the most significant challenge the islands face is the in-migration of mainland workers looking for a job in the lucrative tourism industry. To regulate this migration, the Special Law includes several innovative policy solutions. Among other policies, the law (1) limits temporary admissions to the islands to discourage illegal migration, (2) restricts fishing permits to permanent island residents only, (3) restricts the ability of nonpermanent residents to get jobs on the islands, and (4) places a 1-year time limit on the length of nonresidential employment. In addition, the law includes several incentives for compliance. For example, businesses that hire permanent residents receive tax breaks, effectively establishing an affirmative action for local Galapagueños.

Since the Galapagos Marine Reserve was created in 1989, legislation has expanded its legal protections. The Special Law of 1998 is no exception. The law strengthened the protections of the Marine Reserve by extending the protected area by 40 nautical miles, establishing the participatory management system (Quiroga et al. 2009, 77), and by excluding the industrial fishing sector from the Marine Reserve (Quiroga et al. 2009, 4). There has been much research on the establishment of the Marine Reserve and its impact on protected species. Twenty years after the establishment of the Marine Reserve, scholars found that only 35% of tourists and 27% of fishing industry workers supported its existence (Quiroga et al. 2009, 53). It is perhaps not all that remarkable that after 20 years of the Marine Reserve's existence, only a minority of the tourism and fishing interests actually accepts the concept of the Marine Reserve. Most businesses, tourism and fishing, continue to oppose environmental regulations advocated by most nonprofits.

## Measuring the Impact of the Special Law Through the Perspectives of Local Leaders

Overall, the data suggest that support for the law and ecotourism policy in general is very strong across all leaders surveyed. There was unanimous (100%) support from surveyed leaders on the use of tax credits to reward business for hiring native residents and for the use of quarantine policies designed to prohibit invasive species from entering the islands' habitat. Leaders also extensively support policies to enhance collaborative research on the environment and sustainable development (86–100%, depending on the specific research purposes). Eighty percent support policies to encourage entrepreneurship by supporting local island crafts. When asked about their support for measures designed to limit immigration, 88% of leaders favored the limitation of fishing to residents and 63% supported the limitation of jobs to only permanent residents. We also surveyed leaders about their support for special subsidies. With regard to the Galapagos standard minimum super wage, 100% supported it; 100% also supported the airfare price subsidy. There was strong support among leaders for increasing the required minimum wage for Galapagos teachers (86%). Limiting tourists, both now and in the future, were not features of the Special Law, but we did survey leaders on their views of these two broad policies. We found that 66% support limiting tourists now and 63% support limiting tourists in the future. In sum, we can conclude that the leaders had a positive and enthusiastic level of support for conservation policies.

While we found strong support for provisions of the Special Law of 1998 in general, we found much more disagreement in the leaders' views of the effectiveness of the Special Law itself. The policies viewed as most effective were the teachers' super minimum wage (100%) and limits of fishing activity to residents (100%). A majority of leaders thought that the following policies were effective: airfare subsidies (66%), business credits for hiring Galapagueños (63%), funding for environmental research (63%), and the federal super wage (57%). Leaders were split in their assessment of the effectiveness of limiting jobs to residents (50%) and funding research to protect local natural resources (50%). The least effective policies, as assessed by leaders, were funding sustainable development (28%), funding collaborative research (28%), use of quarantines (25%), and promoting entrepreneurship (14%). In sum, it should be noted that there is a gap between the general support level and the assessed effectiveness for many of these policies. Leaders supported the policies in general, but they felt the implementation was flawed. This phenomenon—the difference between espoused support for certain public policies and support for such policies once they are put into practice—has been well documented by public administration scholars. Research in organizational theory shows that members of an organization—in our case, public and nonprofit sector leaders—often demonstrate a difference between their espoused policy preferences and their support for such policies once enacted and implemented (Argyris 1995; Kaplan 1998).

There are several reasons why leaders might find the policies to be ineffective. First, perhaps the leader does not support the purpose of the policy in general. However, as we saw above, this does not appear to be the case in our research. Leaders voiced support across all broad policy types for almost anything that would balance growth and sustainability. Leaders also expressed that policy effectiveness has been affected by the overall weakness of government institutions. Since 2002, the GNP has had over 14 national directors, which is more than enough to weaken organizational morale and reduce the enforcement and effectiveness of the law (Bassett 2009, 72). A third (and closely related) possibility is the political fragmentation of interests, which leads to a festering of long-lasting conflicts based on economic interests, almost reaching the level of hyperpluralism (such as the conflict between local fishermen and conservationists). There seems to be evidence of this in the literature, but our respondents did not openly voice this concern.

Another possible reason for ineffective implementation is social capital. Theoretically, social capital can either gird up or undercut institutions, depending whether it is bridging or bonding in its effects (Granovetter 1973; Putnam 1995). However, these nascent institutions created in the Special Law were layered on top of the culture of the Galapagos Islands. In a closely knit community, such as this one, social capital may actually prohibit rather than encourage such compliance.[6] A final reason for ineffective implementation is that the old schisms among multiple interest groups, which we previously discussed, have stubbornly remained and

---

[6]What this might mean is that either a resident or, particularly, a leader may be called upon to turn in a relative or close acquaintance because he or she has the wrong immigrant status and has not yet achieved permanent resident status. For example, now a new immigrant must be married to a permanent resident for 10 years before becoming a (legal) permanent resident. As a result, the islands now abound with unofficial residents whose temporary work permits privileges have expired.

inhibited the government from carrying out its regulatory duties. Because the conflicts we discussed previously are rooted in contradictory economic interests, they are unlikely to disappear in the foreseeable future. For example, as recently as July 2011, sharks were found slaughtered and authorities suspected the crime was carried out by local fishers, thus indicating that the fishermen remain militant over refusing to sacrifice their livelihood in the name of conservation (MSNBC 2011).

## A Model Policy for Ecotourism

In the process of interviewing local government and nonprofit leaders, we asked them to give us their vision for a model ecotourism policy for the Galapagos. There was a surprising amount of consensus on what a model policy might look like. The ideas below also echo those published in similar works on this subject (Quiroga et al. 2009, 120–121). These points are meant to represent guiding principles:

1. *Do no harm.* In creating ecotourism policies, the result should not harm conservation efforts. Concurrently, policies should not unravel successful sustainable development efforts.
2. *Involve all actors in decision making.* Inclusivity should be used when developing a tourism model; many actors must be considered, including the local community and government, as well as local, national, and international businesses and nonprofit interest groups. Citizen engagement in Galapagos conservation has been minimal to date, and this must change to ensure a buy-in from residents.
3. *Promote sustainable growth through internal development.* The community should seek self-sufficiency and internal growth over external recruitment of business. In practice, this principle means things like ensuring that the business community uses more locally sourced products. This would be a great improvement over current standards since, for example, many agricultural products used by local restaurants and hotels are imported daily from mainland Ecuador. Part of this principle involves creating a consciousness among the local population for local products.
4. *Give government agencies the tools needed to enforce their mandates.* Government institutions must be given the power to enforce their decisions in a uniform manner. For example, the use of VMS (vessel monitoring systems) to track fishing boats within the Marine Reserve will greatly assist in the prosecution of illegal fishing activities (Quiroga et al. 2009, 120). More generally, the Galapagos National Park needs both the resources and regulatory authority to carry out its job. Also, hand in hand with this is the need for greater revenue for the park.
5. *Uniformly enforce immigration laws.* Immigration laws are necessary to curb the islands' population growth. The Galapagos are unique, in part, because the Special Law limits immigration from mainland Ecuador, within the country, rather than from abroad. However, in our interviews, respondents suggested that the immigration laws are largely ignored.

6. *Galapagos needs a new model of tourism to prevail in the long term.* The islands' current tourism model revolves primarily around cruise ships and short island visits. Cruise passengers engage in "island-hopping" and do not stay long enough to inject substantial resources into the local economy (Taylor et al. 2003, 980). A sustainable tourism model would encourage guests to stay 7–10 days, which is long enough to spend money in the local community. One idea is to tax transient tourists (staying only 2–3 days) at a higher rate than longer-staying travelers.

7. *Encourage cultural change at the community level to embrace sustainable ecotourism policies.* Multiple leaders noted the absence of a conservation culture among local residents. In order for ecotourism policies to be truly successful, the community must support and understand the importance of balanced development and conservation. Currently, there are some efforts to promote education on conservation in the public schools, but more work is needed in this area. In the future, educating the public about conservation should be a core component of ecotourism policy.

## Conclusion

The balance between economic development and conservation remains an ephemeral goal for the Galapagos Islands due to many factors. Our research suggests that the largest barriers in this area include a lack of political will, systematic institutional weakness, and the strong ties of social capital. As one of the survey respondents noted, the Special Law of 1998 was "visionary…but the institutions are not good." Legislation like the Special Law of 1998 is only the first in a series of steps that must be carried out to implement an effective, comprehensive approach to ecotourism policy. At the macro level, the first steps to improvement should begin with institutional-level reforms.

Change is also needed at the micro level. The average Galapagueño resident sorely needs to discover and protect the wonders of their environment. As we have noted, there are some signs of promise here; some of the leaders thought that conservation programs in the public schools were very promising. Creating a culture of conservation that stresses a balance between economic development and environmental preservation must inevitably start at the community level with local residents. Other residents — such as the large immigrant and illegal population — are less likely to be conservation oriented and in fact may resent the government for limiting fishing and other economic pursuits. Our study demonstrates that, even with great leadership support, the Special Law of 1998 in and of itself cannot triumph when it competes with forces innate within the social and political fabric of Galapagos society.

**Acknowledgements** The authors would like to thank Anne Davis, Trevor Fleck, and Sarah Osmer for their background research work on this chapter. We would also like to thank Carmen Huerta-Bapat, whose assistance with translating was invaluable in collecting data for the research findings. Finally, we would like to thank the UNC Center for Galapagos Studies for providing funding for the research project which forms the basis of this chapter.

# References

Argyris C (1995) Action science and organizational learning. J Manag Psychol 10(6):20–26

Barzelay M, Gallego R (2006) From "new institutionalism" to "institutional processualism": advancing knowledge about public management policy change. Governance 19(4):531–557

Bassett CA (2009) Galapagos at the crossroads: pirates, biologists, tourists, and creationists battle for Darwin's cradle of evolution. National Geographic Society, Washington, DC

Bjork P (2000) Ecotourism from a conceptual perspective, an extended definition of a unique tourism form. Int J Tourism Res 2:189–202

Blamey RK (2010) Ecotourism: the search for an operational definition. J Sustain Tourism 5:109–130

Donohoe HM, Needham RD (2006) Ecotourism: the evolving contemporary definition. J Ecotourism 5:192–210

Einarsdottir K, Graburn N, Palsson G (2012) Discrepancies between defined and actualized ecotourism: bridging the gap between theory and reality. LAMBERT Academic Publishing, New York

Epler B (2007) Tourism, the economy, population growth, and conservation in Galapagos. The Charles Darwin Foundation, Puerto Ayora, Galapagos

Fennel DA (2001) A content analysis of ecotourism definitions. Curr Iss Tourism 4:403–421

Fennel DA (2003) Ecotourism: an introduction. Routledge, London

Fennel DA, Dowling RK (2003) Ecotourism policy and planning. CABI Publishing, Cambridge

Granovetter MS (1973) The strength of weak ties. Am J Sociol 78:1360–1380

Galapagos National Park (2011) Valores Recandados for Ingresos de Turistas. Ministerio del Ambiente. Ecuador

Hearn A, Martínez P, Toral-Granda MV, Murillo J, Polovina J (2005) Population dynamics of the exploited sea cucumber *Isostichopus fuscus* in the Western Galapagos Islands, Ecuador. Fish Oceanogr 14(5):377–385

Jácome C (2008) Subsidios en el sector energético insular. Informe Galápagos 2006–2007. FCD-PNGINGALA, Puerto Ayora, Galapagos

Kaplan RS (1998) Innovation action research: creating new management theory and practice. J Manage Accounting Res 10(10):89–118

Knott J, Miller G (1987) Reforming bureaucracy: the politics of institutional choice. Prentice Hall, Englewood Cliffs, NJ

MacDonald T (1997) Conflict in the Galapagos Islands: analysis and recommendations for management. Report from January 1997. Harvard University, Cambridge, MA

MSNBC (2011) 'Marine massacre': hundreds of dead sharks buried at sea. http://www.Msnbc.com. http://www.msnbc.msn.com/id/43882266/ns/world_news-world_environment/#.T3XSbdVBpJ4. Accessed 1 Apr 2012

North D (1990) Institutions, institutional change, and economic performance: political economy of institutions and decisions. Cambridge University Press, New York

Page S, Dowling (2002) Ecotourism. Pearson Education. Harlow, UK

Putnam R (1995) Bowling Alone: America's Declining social capital. J Democracy 6:65–78

Quiroga D, Mena CF, Suzuki H, Guevara A (2009) Galapagos marine area socioeconomic and governance assessment. Unpublished report. Conservation International and Universidad San Francisco de Quito, Quito, Ecuador

Taylor E, Dyer G, Stewart M, Yunez-Naude A, Ardila S (2003) The economics of ecotourism: a Galapagos Islands economy-wide perspective. Econ Dev Cult Change 51:977–997

Taylor JE (2006) Ecotourism and economic growth in the Galapagos: an island economy-wide analysis. Agriculture and Resource Economics working paper series. Paper No.06–001. University of California at Oavis

Valentine PS (1993) Ecotourism and nature conservation: a definition with some recent developments in Micronesia. Tourism Management 14:107–115

Watkins G, Cruz F (2007) Galapagos at risk: a socioeconomic analysis of the situation in the archipelago. Charles Darwin Foundation, Puerto Ayora, Galapagos

World Wildlife Fund (2012) The Galapagos: the people. World Wildlife Fund. http://www.worldwildlife.org/what/wherewework/galapagos/people.html. Accessed 1 Apr 2012

# Chapter 8
# People Live Here: Maternal and Child Health on Isla Isabela, Galapagos

Rachel Page, Margaret Bentley, and Julee Waldrop

## People Live Here!

Although famed for the riches of the flora and fauna on the islands, the Galapagos Islands have a long history of human presence. Dating back as far as the Incans, there is evidence of successful trips between the islands and the mainland around the late 1400s. The first documented case of an individual living on the archipelago is credited to Patrick Watkins in 1807. The islands were a popular location for sea-faring voyagers to restock food (giant tortoises and sea lions), as did Darwin's boat, *HMS Beagle*, in 1835 (Stewart 2006). Over the decades, there were several attempted settlements, and in 1893, a colony was founded on present-day Isabela Island (Constant 2006). The islands have also served as refuge for pirates, sailors crossing the Pacific, and wanderers for hundreds of years, with the culmination of the present-day inhabitants and the creation of a recognized Ecuadorian province, with a provincial government located on San Cristobal Island (Bassett 2009).

The archipelago consists of 14 volcanic islands, four or which are currently populated: Santa Cruz, San Cristobal, Isabela, and Floreana. The capital is Puerto Baquerizo Moreno on San Cristobal. Ninety-seven percent of the geographic area of the Galapagos is a designated national park, with only a small and increasingly populated area available for human habitation (Parque Galapagos 2011). Each year the number of tourists visiting the Galapagos Islands increases,

R. Page, M.P.H. • M. Bentley, Ph.D.
Gillings School of Global Public Health, University of North Carolina at Chapel Hill, Chapel Hill, NC, USA

J. Waldrop, D.N.P. (✉)
School of Nursing, University of North Carolina at Chapel Hill, Chapel Hill, NC, USA

College of Nursing, University of Central Florida, Orlando, FL, USA
e-mail: julee.waldrop@ucf.edu

S.J. Walsh and C.F. Mena (eds.), *Science and Conservation in the Galapagos Islands: Frameworks & Perspectives*, Social and Ecological Interactions in the Galapagos Islands 1, DOI 10.1007/978-1-4614-5794-7_8, © Springer Science+Business Media, LLC 2013 with approximately 173,287 in 2010 arriving primarily by plane and then touring in a variety of sizes of boats, ferries, ships, and yachts. Staying in hotels on the islands is increasingly popular among tourists, providing opportunities for employment for the local population while at the same time creating environmental pressures on a fragile ecosystem.

The Galapagos Islands have experienced significant population growth, increasing more than 300% in the past few decades. The 1990 census marked the population at 9,735, whereas the 2010 census listed the official total population at 25,124 residents. Santa Cruz has the largest population at 15,393, with 2,256 on Isabela Island (Instituto Nacional de Estadistica y Censos 2011). Floreana has only 120 residents (last counted in 2006). The population is doubling every 11 years, and it is estimated that there will be 40,000 people on the islands by 2014 (Bureau of Statistics of Ecuador 2006). Because of this rapid growth, the Special Law of the Galapagos was passed in 1998. This law placed restrictions on migration to the islands by limiting residency to only those living on the islands in 1998. It is estimated that 20% of the residents in the Galapagos do not have government permission to live there (Patel 2009).

Higher wages (up to 70%) and better living conditions on the islands, compared to on the mainland, continue to fuel this recent wave of immigration. In Ecuador, 46% of the population falls below the poverty line (United Nations Development Program 2009). Although specific figures for residents of the Galapagos are not available, it is likely that poverty levels are much lower. However, despite the economic lure, migration can disrupt family relationships, social networks, and access to resources in the new location for migrants and locals alike (Acosta et al. 2006).

According to the Charles Darwin Foundation (CDF), a local nongovernmental organization active in the development and conservation of the Galapagos, the "economic growth has resulted in unsustainable population growth, socioeconomic stratification, civil unrest, strained public services and infrastructure, an increase in the number of invasive species, and a number of conflicts with conservation goals and authorities" (Epler 2007). Surprising to many is the fact that there is a scarcity of fresh water in the Galapagos. This, in addition to a lack of wastewater treatment and sanitation facilities, greatly impacts health conditions (Walsh et al. 2010). Due to increasing economic and population growth, the geographic isolation of the islands, and agricultural planting restrictions for the preservation of indigenous flora, both food security and food quality have direct health impacts upon the residents of the islands. In addition, the strain placed upon utility infrastructure results in water contamination, affecting the human residents as well as sea life in the surrounding waters.

Although little has been published about the health situation on the islands, Galapagos residents face several challenges to protecting their nutrition and health status. There is no mention of the health in the comprehensive report by the Charles Darwin Foundation (Epler 2007) and a search on PubMed from 1995 to April 13, 2012 revealed only one research study (Walsh et al. 2010) published in English with references to human health in Galapagos.

The goal of this study was to better understand the participants' personal health concerns and their perceptions of the health of their young children. Interviews focused on: diet, food preparation, shopping patterns, knowledge of appropriate feeding and dietary needs, lifestyle and physical activity patterns, health seeking behavior, and reasons for migrating to the islands.

## Data Collection Methods

We have collected multiple sources of data for this paper, including observations, informal interviews, a short survey, and in-depth home interviews with mothers of children under 2 years of age. In 2009, one of the authors (Waldrop) visited Isabela, interviewed key informants and observed facilities related to the availability of health services. Early in 2010, a pen and paper survey was distributed to a convenience sample of adult students taking English classes and living on Isabela. To participate in the survey, the student had to be the parent of at least one child under 5. The survey contained 32 questions, and 18 respondents answered most of the questions. Table 8.1 describes the characteristics of this sample. The results of this survey were used to inform the development of questions for in-depth, semi-structured interviews conducted in summer 2010 by the other two authors (Page and Bentley). The interview data are primarily qualitative but include the collection of some quantitative questions and measures, such as anthropometry of the mother and child, a depressive symptoms instrument, and body size preference data. The interview guide consisted of primarily open-ended questions; however, some sections utilized short previously validated tools for specific topics that were scored independently (The Center for Epidemiologic Studies Depression Scale CES-D) (Radloff 1977). Potential participants were identified through referrals from the nurse in the government clinic (*subcentro*). Twenty interviews were conducted in Spanish with mothers of children of less than 5 years of age in their homes (18) or in the workplace (2).

**Table 8.1** Participant characteristics (surveys)[a]

| Characteristics | Mean (range) | $N$ (%) |
|---|---|---|
| Sex (female) | | 15 (83) |
| Age (years) | 28.8 (21–38) | |
| Time on island (years) | 13.8 (1–34)[b] | |
| Education: high school or more | | 18 (100) |
| Employed full time | | 6 (33) |
| Employed part time | | 6 (33) |
| Unemployed | | 6 (33) |
| Married or union libre[c] | | 15 (83) |

[a]$N = 18$
[b]$N = 13$, excludes five who have lived on the island their whole lives
[c]Civil union

**Table 8.2** Maternal characteristics (interviews)[a]

| Characteristics | Mean (range) | N (%) |
|---|---|---|
| Age (years) | 28.5 (19–37) | |
| Time on island (years) | 7.75 (1–17)[b] | |
| Education: high school or more | | 13 (65) |
| Religion: Catholic | | 14 (70) |
| Employed full time | | 8 (40) |
| Employed part time | | 3 (15) |
| Unemployed | | 9 (45) |
| Married or union libre[c] | | 18 (90) |

[a]$N = 20$
[b]$N = 16$ (excludes four who were born on island)
[c]Civil union

The index child was designated as the youngest child under 5 and over 6 months of age in each household. Each interview lasted approximately 1 h and was audio recorded. All participants provided informed consent in Spanish, in writing and orally. Table 8.2 describes the characteristics of this sample.

In addition to the interviews with mothers, contact was made with local medical professionals (including all active doctors, the director of the health center, and the lead nurse). Informal interviews were conducted with local government officials, such as the mayor, local pharmacists, directors of local nongovernmental organizations and local school officials.

## Data Analysis

The qualitative data from the in-depth and key informant interviews were coded and analyzed using a software program, ATLAS.ti. Codes were generated from the interview guide and additional codes were added based upon reading the transcripts. Display matrices (Miles and Huberman 1994) were generated that summarized specific categories of data, such as "perceived health problems" or "water quality." Survey data were quantified into response percentages for each question.

## Life on the Island

Isabela Island is a quiet corner of the world. There are two ways on and off the island—ferry boat or a 9-seater plane—though most residents travel by boat when they leave or return to the island. The boat ride to Isabela from Santa Cruz is approximately 2 h across the high sea. Upon arrival, one can hire one of a handful of "white trucks" (a dollar for a ride anywhere in the main part of the city), or one can walk the mile into town. Most of the roads consist of sand and volcanic pebbles and, in certain areas, are shaded by local vegetation, non-indigenous palm trees, and scattered volunteer houseplants.

The principal city on Isabela, Puerto Villamil, has a central plaza, surrounded by local government offices, a variety of stores, restaurants, and the local Catholic church. The island has an open and friendly atmosphere, with locals passing by with salutations and ease.

When asked, "what is it like to live on the island?" the majority of women interviewed (13 or 65%) used the Spanish phrase "tranquilo," or "calm, peaceful, relaxing, and tranquil." This response was typically followed with statements about personal security on the island, such as:

> La tranquilidad. (it's peaceful, calm)... it's not dangerous, it's not too expensive... the natural environment... No one lives a hectic or busy city life. Where you are always running somewhere then running somewhere else. No, here, it's more peaceful, calm (tranquilo). (married woman, 36)

Another interview question asked women, "why do you live on the island?" The most frequent responses made reference to a spouse's need to secure work (4) and because, in contrast to the mainland, the island provided a safe environment for their children to play outdoors (4). One mother expressed concern for her child when she visits the mainland:

> When we are there (Guyaquil), she gains a lot of weight, and I think it's because there isn't anything to do, you can't do anything, because there aren't any parks, and it's always dangerous... Here we have the freedom to run, to go out, to play, go to the beach, even though things can be dangerous, its way less dangerous than it is in Guyaquil. (married woman, 28)

Economic opportunities provided an incentive to move to the Galapagos, and relatively safe communities encouraged migrants to stay. However, despite these benefits, the rapid increase in population has not been matched by supporting infrastructure, and the participants have some serious concerns about problems that impact their health and well-being.

## Common Problems on the Island

When survey respondents were asked to list the top three health problems, they noted the following: lack of a hospital or emergency services, inexperienced health personnel, and a shortage of specialists and medications (13/18). One woman stated:

> I believe that knowledgeable doctors are instrumental. There is a lack of all the necessary tools to medical care, medicines, most of all. So that, for example, you go to the doctor and (he) says buy yourself this medicine, but one goes to look to pharmacies and it is not there. One must often send to Santa Cruz or elsewhere. Imagine an emergency, what happens? The patient dies because there is no way. So I think that's quite necessary here. (married woman, 37)

A few women surveyed noted problems associated with poor diet, such as cardiovascular disease and cholesterol (3/18). Another woman interviewed reported:

> I understand that the majority of the people here suffer much from cholesterol…, triglycerides, hypertension, diabetes, for the food here, people eat meat and pork and fries, things like that with a lot of fat in the food. For me, I do not like and I have to take care of my daughter because I see that she is chubby. (married woman, 31)

The three most serious problems survey respondents listed were lack of emergency care (12/18), diarrheal illness (12/18), and lack of potable water and sewage contamination (6/18). One woman said:

> My daughter suffers from asthma and needs a spray (ventolin) and it is not always available here. This aspect we can control but in an emergency… There are many who have died on the dock from medical inattention or lack of transportation. (married woman, 28)

## Access to Healthcare

### Medical Infrastructure on Isabela Island

As in most of rural Ecuador (Lopez-Cevallos and Chi 2009) on Isabela, the Ministry of Health provides services via a recent graduate from an Ecuadorean medical school assigned through the 1-year obligatory rural medical services program in the *subcentro* (health center). The *subcentro* has 8–10 rooms used for patient care, a lab, two storage areas for medications and vaccinations, and a reception area. There are two rooms that could be used to provide hospital-like care, but there are not enough nurses on the island to provide this service. There is also a room that could be used for surgery which is equipped with oxygen, suction, and rudimentary anesthesia equipment; however, without an anesthesiology provider, this is also not utilized. One room is used to deliver babies, but there are no fetal monitoring capabilities, except for a handheld Doppler (ultrasound) that allows only auditory assessment of the fetal heart rate. A 15-year-old ultrasound machine can also be used for in utero evaluations. This clinic performs 1–2 deliveries a month, but if a women can afford it, she will typically travel to Guayaquil before her due date to deliver her baby at a hospital that can provide emergency intervention if necessary. This clinic has a radiograph machine, but no one trained to take the X-rays. The lab can perform blood counts, chemistries, and microscopic analysis on blood samples. There are also a centrifuge and a microscope for evaluation of urine. The clinic has the ability to test for and treat sexually transmitted diseases. Prenatal care and immunization clinics are also provided.

The primary physician interviewed at the clinic believes there has been an increase in adults with diabetes and hypertension on the island. She also reported that there are seven brothels on the island and many other sex workers who are waitresses in the bars. Chlamydia and gonorrhea are a problem, and it is possible there are HIV cases because there are some on Santa Cruz Island (personal communication from hospital doctor on Santa Cruz to Bentley 2010). The clinic provides IUDs (Copper T) and oral contraceptive pills, as well as condoms when available.

A private municipal clinic, the *policlinico*, also provides sporadic physician services for a fee or in a health maintenance organization style with monthly $10 payments. This is a smaller but newer facility with only one patient exam room. There are rooms for lab equipment, childbirth, and surgery. The lab has newer equipment than the *subcentro* and can perform similar tests. The labor room is equipped with a fetal monitoring system, but no ultrasound is available. The room for surgery lacks any equipment. There is no potential for X-rays here. Approximately once a month, this facility hosts "campaigns" where physicians come over from the mainland with all the equipment needed to provide a week's worth of care for patients. There is also one active privately practicing physician who also provides health services.

Since there is no ability to perform surgical procedures under general anesthesia, all residents and tourists within the archipelago in need of these services must travel to Santa Cruz (for minor procedures) or the mainland over 600 miles away for major procedures, such as cesarean births. However, a large, modern, and better-equipped hospital on San Cristobal Island is currently under construction with the expectation of providing surgical, obstetrical, gynecological, dental, and preventive health services to all residents and visitors within the archipelago (Basantes, 2011, personal communication).

## Childbirth on Isabela Island

Of the mothers interviewed, only two chose to have their children on Isabela. The survey participants were not asked about the birthplace of their children, but of the five participants who reported living their whole lives on Isabela, three reported being born on the mainland. The survey and interviews did not specifically address the mother's decision of *where* to birth her children; however, it should be noted that minimal health services pose an increased risk for women who wish to have their children on Isabela, as well as an increased financial burden for women who must travel to the mainland to seek neonatal healthcare facilities. Although there are no official statistics on infant deaths during delivery on the island, the day before one of the authors arrived on Isabela in 2009, an infant had just died during delivery (Sanchez, 2009, personal communication).

## Perceptions of Child Health and Medical Care

Several of the mothers interviewed (17/20) stated that they believed that their child's health was "good"; when asked to elaborate, a common explanation involved the local doctors stating that their child was "good" or "healthy." While the opinions of medical professionals were considered, many expressed concerns over the lack of well-trained medical professionals. One woman responded to the question, "how do you see the overall health of your child?" by explaining her doubts about the local medical providers:

> Yea... I can't tell you 'good,' 'good,' because I don't know what illnesses can do, I don't know... I have my doubts... you know, that baby has a little growth, and here, the doctors say that it is a node, but that little ball, every once in a while it grows a little more, and they didn't tell me anything..., I would like to take him outside (mainland) so that I can get an ultrasound, I think that's what it's called, to see if it is what the doctors here say it is..., or to see if it is something else, sometimes, you know, they can be tumors... (co-habiting woman, civil union, 29)

Even though the women did not completely trust their local doctors or have confidence in their medical training, most felt they had no other options for obtaining medical advice or care.

Limited medical services and constant staff changes within these services affected the decision making of mothers concerning where to receive medical attention. Half of those interviewed identified the *subcentro* as their primary medical location. However, obtaining local free services, long waits, changing staff, and inaccurate diagnoses were among several deterrents. Medical attention provided at the *subcentro* is commonly performed by a physician who has just completed medical school (as described above) with little to no oversight. One woman stated, "I go there because it is the only choice I have." All other medical services on the island are private and therefore have associated costs. Some of the women mentioned that if the *subcentro* was closed (weekends or nights), then they would have to take their child to get private medical attention.

One quarter of the women interviewed preferred to receive medical attention (annual checkup, etc.) for their children from pediatricians on the mainland, whenever their finances allowed them to make this journey. One woman explained that she prefers to call her pediatrician on the mainland and correspond via email in order to receive basic medical advice for her child. Families with limited disposable income are at a significant disadvantage for medical treatment on Isabela; their only option is to receive services at the *subcentro*. Five women clearly stated that financial reasons were why they did not seek medical attention at the private *policlinico* (a charge of $2 USD is the average cost for a child consultation at a private clinic on Isabela). "Here, at 4 p.m., they close the hospital (subcentro). If there is an accident, you have to go find a doctor and you'll die before you find one" (married woman, 24).

Survey respondents (16/18) echoed these sentiments, reporting that they did use the local healthcare providers for medical attention and advice, with only two mothers reporting that they would call or travel to the mainland to see a pediatrician when their children were ill. However, parents also reported using a variety of remedies for common illnesses, such as a cough or cold, ranging from vitamin C to probiotics and other over-the-counter products. Most felt that there were adequate over-the-counter medicines available but prescription drugs were often in short supply or unavailable.

Eighty-three percent of those surveyed reported that they left the island for medical care at some point. The primary reason was for childbirth. Other reasons why care was sought on the mainland included no emergency care or X-rays, lack of specialists or experts, and a lack of confidence in the providers and services available. When asked what services they would like to see available on the island, pediatric care, obstetrics, and gynecology were at the top of the list. A variety of other specialists were also listed.

## Food and Nutrition

Food arrives on Isabela by boat or airplane, or is grown in the highlands. Produce and some animal source foods grown in the highlands are brought to town twice a week to sell in the local Saturday market during the dry season and, as weather and roads permit, in the rainy season. Almost all of the markets allow clientele to purchase products on a line of credit that the families pay back at a later date (on payday or in increments).

About half of the women interviewed reported that they were the primary food purchasers within their households, and 4/20 said that their husband was the primary shopper. For 4/20, husband and wife shared the food purchasing power/responsibility. A small number depended on a mother/mother-in-law to purchase food for the family. Residents were also asked where they usually shopped for food. Most listed the many commercial vendors in Puerto Villamil (16/18 surveyed) and about half of these also shopped at the local Saturday market.

The majority of respondents surveyed or interviewed reported that there were sometimes or always shortages or a scarcity of the foods they would like to purchase. Most listed vegetables as the food they would like to buy that was frequently not available. Other items not available were grains, varieties of cheeses, and meat products. The cost of food was sometimes a barrier (6/20 interviewed). One family reported importing foods with a long shelf life (rice, sugar, oil) from the mainland as a way to economize, but that this option was not available to those without partners on the mainland and/or the financial means to complete such transactions. Sometimes food arriving from the mainland was already spoiled (2/20). One islander said:

> I'll tell you the truth… you can't provide 100% nutrition, because here, you can't have 100% nutrition, not for the children or for the adults, because the systems, it's really, really difficult to get food here. If I wanted to give an apple to my son, it (the apple) has to come 4, 5, sometimes 8 days in a boat, and the boat, if you could see one of these boats, where the products come from, it is in horrible conditions. The fruits are mishandled, sun burnt and exposed to heat, they arrive soaked… (Husband of a mother interviewed, 37)

Despite complaints of shortages, most felt that they could procure a "quality diet" although the variety of foods, especially vegetables, was limited. The same thoughts were echoed when asked about the quality of their children's diet. Most parents reported that they were able to give their children a balanced diet, although, at times the lack of availability, variety and/or high cost of vegetables was a barrier.

## Water and Sanitation Issues

The municipality provides the principal water source for Puerto Villamil. The only treatment provided is filtration through a series of sieves in order to remove pebbles and rocks (Walsh et al. 2010). On Isabela, the primary water source is located near the northern part of town, and if one follows the same road out of town and up the mountain, the island's dump will be found. Several people expressed concerns about waste from the dump leaking out and entering the already vulnerable water source.

Several women interviewed (12/20) identified water as a primary area of concern and one of Isabela's greatest challenges. One quarter noted that the water was undrinkable and that payment was required for water access or treatment. Several noted that the contaminated water caused infections (8/20), especially vaginal infections among women and skin infections among children. These same concerns were also highlighted by the survey participants. They reported diarrheal illness (12/18) and the lack of potable water and sewage contamination (6/18) as two of the top three most serious problems on the island. In addition, there were several reports of children getting sick from accidently swallowing untreated water.

> I wish you could take a sample of the tap water to a lab. It isn't even acceptable, even to bathe with. And this water has to be used to prepare foods, and sometimes (wash) the children. So, look, that's what we are suffering from, and that is what we are hoping for, and wishing that they can treat it (the water). (married woman, 37)

When those surveyed were asked if they could change one thing about living on the island, the second most common response was "clean water." One stated "The water, to have potable water, it is the fundamental problem and it is very serious for health."

## Additional Health Issues

The 20 women and children interviewed had their weights and heights measured in their homes. Overweight and obesity were common: nine women were overweight (BMI greater than 25) and five women were obese (BMI over 30) (Table 8.3). According to the World Health Organization (2008) $z$-score definitions, ten children had weight for length/height or BMIs in the normal range, six children were at risk for overweight ($z$-score above 1), one child was overweight ($z$-score above 2) and one was obese ($z$-score above 3). In this sample of children, there was also one child whose weight for age was wasted ($z$-score below $-2$) and one who was severely wasted ($z$-score below $-3$). The severely wasted child was also severely stunted ($z$-score below $-2$) according to length for age.

When queried about their perceptions and preferences for their own body size, the majority of mothers preferred a smaller body size than their perceived current body size, both for personal and for health reasons. However, their perceptions and preferences differed when reporting on their children. The mothers preferred their

**Table 8.3** Women's anthropometry (interview)[a]

|               | Mean (range)       | N (%)   |
|---------------|--------------------|---------|
| Height (cm)   | 154 (142–164.7)    |         |
| Weight (kg)   | 65.25 (47.3–89.1)  |         |
| BMI (mean)    | 27.3 (20.3–37.4)   |         |
| Normal        |                    | 6 (35)  |
| Overweight    |                    | 9 (45)  |
| Obese         |                    | 5 (20)  |

[a]$N=20$

children to be larger than they currently perceived them, either for personal reasons and/or because they felt that a larger child reflects a healthier child.

The women interviewed were also screened for symptoms of depression using the CES-D. This is a screening tool for the presence of symptoms associated with depression. Scores ranged from 3 to 30 (highest possible score=60); the higher the score, the higher the endorsement of depressive symptoms. Five women (25%) scored above 15, which is indicative of a significant level of psychological distress. In the normal population it is expected that approximately 20% will score above 15 (Radloff 1977).

When asked to describe one thing they would change about life on Isabela, the most frequent response was to increase available activities and educational venues. This was expressed as wishing for opportunities to finish or continue their studies. As one person said, "if I had not had to abandon my early studies I would be a professional today." Another respondent could not choose just one thing and stated:

> If I could change medical care, I mean so that you would not have to go to Guayaquil or Quito to solve something because if I could (have) done the same right here. Education would also be something important because this (here) does not meet the demands of the modern world.

## Conclusion and Discussion

This exploratory study about women's experiences living on Isabela Island, and their concerns about the health and nutrition of their families, identifies several problems that could be addressed by the local and national governments. A major concern is the lack of consistent quality medical care. The lack of any specialists, such as pediatricians and gynecologists, is mentioned repeatedly. Another key problem for locals and tourists alike is the limited access to emergency services. Increases in population and in the number of tourists who visit the island make this a significant issue.

Worries about the lack of fresh drinking water and health problems associated with contaminated water, such as fungal skin infections, parasite-associated diarrhea, and vaginal and urinary tract infections support the findings of Walsh et al. (2010).

Lack of access to fresh water for drinking, cooking, or bathing compromises parents' ability to ensure healthy growth and development of their children, particularly for those younger than 5 years old that are the most vulnerable.

Many are also concerned about the limited access to locally produced fresh foods, especially vegetables. They dislike dependence on imported, expensive food and the irregular delivery of food and supplies via boats and ships. Many noted concerns about the high cost of healthy foods. The increased availability of highly processed foods that have a longer shelf life but are also less nutritious contributes to obesity here as elsewhere in the world.

We also found that classic characteristics of the nutrition transition currently exist on Isabela Island (Popkin 2006). We identified the presence of "dual burden households," defined as the coexistence of individuals who exhibit signs of both under- and overnutrition within the same households (Doak et al. 2002; Popkin 2006; Waters 2006). For example, in very young children (under 5 years of age), the lack of adequate nutrition and feeding, coupled with high rates of gastrointestinal infection, may result in stunting and poor development. Among adults, overweight and obesity are prevalent and similar to those in urban Ecuador Bernstein 2008). Obesity and overweight in Ecuador, as elsewhere, are associated with high prevalence of chronic diseases, such as diabetes, cardiovascular disease, and hypertension (Bernstein 2008).

However, despite the many reported problems, most women appreciate the positives of living on Isabela Island, such as the tranquility and the perceived safety for their children, and the opportunity for employment and higher income.

As we noted in the title of this chapter, people *do* live in the Galapagos and they will continue to provide the services and products that are in demand by increasing population growth and tourism. Access to adequate health services, food, and water are the basic requirements for human health and well-being. The residents of the Galapagos deserve no less.

# References

Acosta P, Calderón C, Fajnzylber P, López H (2006) Remittances and development in Latin America. World Econ 29:957–987

Bassett CA (2009) Galapagos at the crossroads. National Geographic Society, Washington, DC

Bernstein A (2008) Emerging patterns in overweight and obesity in Eduador. Pan Am J Public Health 24(1):71–4

Instituto Nacional de Estadistica y Censos (2011) Censo de poplacion y vivienda.http://www.inec. gob.ec /cpv. Accessed 4 April 2012

Constant P (2006) Galapagos: a natural history guide. Odyssey, Hong Kong

Doak C, Adair L, Bentley ME, Fengying Z, Popkin B (2002) The underweight/overweight household: an exploration of household sociodemographic and dietary factors in China. Public Health Nutr 5:215–222

Epler B (2007) Tourism, the economy, population growth, and conservation in Galapagos. Charles Darwin Foundationm, Galapagos

Lopez-Cevallos DF, Chi C (2009) Health care utilization in Ecuador: a multilevel analysis of socio-economic determinants and inequality issues. Health Policy and Planning 25:209–218

Miles MB, Huberman AM (1994) Qualitative data analysis: an expanded sourcebook, 2nd edn. Sage Publications, Inc, Thousand Oaks, CA

National Bureau of Statistics of Ecuador (2006) 2006 Galapagos census. http://www.inec.gov.ec/web/guest/home. Accessed 14 Aug 2011

Parque Nacional Galapagos Ecuador (2011) http://www.galapagospark.org. Accessed 3 Oct 2011

Patel T (2009) Immigration issues in the Galapagos Islands. Council on Hemispheric Affairs. http://www.coha.org/immigration-issues-in-the-galapagos-islands/. Accessed 14 Aug 2011

Popkin B (2006) Global nutrition dynamics: the world is shifting rapidly toward a diet linked with noncommunicable diseases. Am J Clin Nutr 84:289–298

Radloff LS (1977) The CES-D Scale: A self-report depression scale for research in the general population. Applied Psychological Measurement 1:385–401

Stewart PD (2006) Galapagos: the islands that changed the world. BBC Books, London

United Nations Development Program (2009) Human Development Report 2009. United Nations Development Programme, New York, NY

Walsh S, McCleary A, Heumann B (2010) Community expansion and infrastructure development: implications for human health and environmental quality in the Galapagos Islands of Ecuador. J Lat Am Geogr 9:159

Waters WF (2006) Globalization and local response to epidemiological overlap in 21st century Ecuador. Glob Health 19:8

World Health Organization (2008) Training course on child growth assessment. World Health Organization, Geneva. http://www.who.int/childgrowth/training/module_c_interpreting_indicators.pdf. Accessed 12 Oct 2011

# Chapter 9
# Characterizing Contemporary Land Use/Cover Change on Isabela Island, Galápagos

Amy L. McCleary

## Introduction

Areas within and adjacent to human settlements in the Galápagos Islands have undergone significant changes in the last three decades. Humid upland areas on inhabited islands have been transformed by introduced and invasive plants and animals (Walsh et al. 2008; Henderson and Dawson 2009; Watson et al. 2009; Guézou et al. 2010). Coastal communities have become more urbanized with the expansion and densification of buildings and the development of transportation infrastructure to support growing local and tourist populations (Walsh et al. 2010; Gardener and Grenier 2011; Cléder and Grenier 2010).

Timely and accurate information about land use/cover change is invaluable for guiding land management and conservation decisions in and around protected areas like the Galápagos National Park (GNP). For example, understanding current patterns and processes of land use/cover change is key for the development of site-specific management plans (Brandt and Townsend 2006) and conservation strategies (Alo and Pontius 2008). However, such assessments are often difficult to conduct in remote areas of developing countries because of limited data, financial constraints, and issues of accessibility (Brandt and Townsend 2006). Such is the case in the Galápagos Islands where information about current land use/cover and past trends is lacking in spite of the rapid changes taking place in the archipelago (Gonzalez et al. 2008).

Land use and land cover information for Galápagos is often incomplete and outdated. The first archipelago-wide maps of land use in Galápagos were produced by the National Institute of Galápagos in 1987 as part of an effort to inventory features of the natural environment (INGALA, PRONAREG, ORSTOM 1987). However, land use maps were not produced for two of the four inhabited islands, Isabela and

A.L. McCleary (✉)
Department of Geography, Center for Galápagos Studies,
University of North Carolina at Chapel Hill, Chapel Hill, NC, USA
e-mail: alnorman@email.unc.edu

S.J. Walsh and C.F. Mena (eds.), *Science and Conservation in the Galapagos Islands:*       155
*Frameworks & Perspectives*, Social and Ecological Interactions in the Galapagos Islands 1,
DOI 10.1007/978-1-4614-5794-7_9, © Springer Science+Business Media, LLC 2013

Floreana. More recently, The Nature Conservancy, with cooperation from several Ecuadorian government agencies, produced a series of land use/cover maps of the Galápagos using data collected in 2000 (TNC and CLIRSEN 2006). The lack of data for some islands and the coarse nature of existing maps have hampered efforts to quantify changes in vegetation (Villa and Segarra 2010) and human-mediated degradation (Watson et al. 2009) on inhabited islands.

Remote sensing and image interpretation have become standard approaches for mapping land use/cover. Remotely sensed imagery can not only cover large spatial extents but can also capture information for features of small grains and extents, particularly with the increased availability of high spatial resolution data products. Image interpretation and GIScience methodologies include automated approaches for mapping that are efficient and easily repeatable, which can reduce the costs associated with in situ data collection. Further, remote sensing can provide information on areas that are difficult to access because of their isolation, difficult terrain, or other constraints (e.g., private land restrictions).

The goal of this chapter is to provide an improved understanding of contemporary land use/cover dynamics in the Galápagos Islands by drawing on a case study of southern Isabela Island. The study area, which encompasses the rural community of Santo Tomás and an area within the adjacent Galápagos National Park, is an important site for exploring landscape change in the archipelago. The humid upland areas are important places where agricultural activities and some of the first human settlements in Galápagos coincide with sites of high biodiversity (MacFarland and Cifuentes 1996). The objective is to first explore the dynamics of land use/cover using a combination of remote sensing data and methods, and field observations. An object-based classifier is applied to high spatial resolution satellite images from 2004 (QuickBird) and 2010 (WorldView-2) to generate land use/cover maps of the region. The dominant cover classes are quantified in each period, and from–to change matrices are calculated to determine the degree of change and major transitions between 2004 and 2010. In addition to general classes representing the most common land use/cover types identified during fieldwork in 2008 and 2009 (barren, built-up, dry pasture/grass, crops/pasture/grass, lava, soil, and forest/shrub), the distributions of two invasive plants are also mapped—common guava (*Psidium guajava* L.) and rose apple (*Syzygium jambos* L). Second, descriptive statistics derived from secondary data sets that include two population censuses (2001 and 2010), an agricultural census (2000) and a living standards survey (2009), as well as information from interviews with local residents (conducted in 2008) are leveraged to contextualize the land use/cover results.

## Study Area: Santo Tomás , Isabela Island

This study is centered on the rural community of Santo Tomás (52 km²) and an adjacent area within the Galápagos National Park (37 km²). This site is located along the southeastern slope of Sierra Negra Volcano on Isabela Island, between

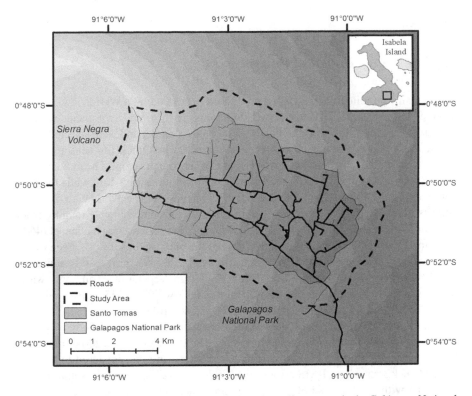

**Fig. 9.1** The study area encompasses Santo Tomás and an adjacent area in the Galápagos National Park in southern Isabela Island

0°47′–0°53′ S and 91°06′–90°59′ W (Fig. 9.1). The climate is semi-arid and subtropical with two distinct seasons—a rainy, warm period from December to June and a dry, cool episode from July to November (Collins and Bush 2011). The relief of the study area is gently sloping, with isolated hills formed by parasitic cones. Elevation ranges from 80 to 1,040 m and slope angles range from 0 to 42°. Vegetation in the site is divided into two commonly recognized zones that progress upward in elevation: (1) the transition zone composed primarily of evergreen plants and (2) the humid zone where introduced vegetation dominates areas once occupied by endemic Scalesia and fern–sedge communities (Wiggins and Porter 1971; Froyd et al. 2010).

Santo Tomás (officially, Tomás de Berlanga) is a community of less than 200 persons that has been continuously inhabited since the late 1890s. It is characterized by smallholder agriculture, agroforestry, and small-scale livestock production. An increasing amount of land within the community is no longer actively managed or given any particular use, which has led to the spread of plants introduced for cultivation (Walsh et al. 2008). The national park, in contrast, strictly controls access to protected areas and limits activities within its boundaries in order to protect native and endemic flora and fauna.

# Methods

## Satellite Image Data and Preprocessing

A QuickBird satellite image acquired on 22 October 2004 and a WorldView-2 image acquired on 23 October 2010 were used in this analysis. The images coincide with the period of peak agricultural production from July to December and were selected based on the availability of nearly cloud-free satellite data. The QuickBird sensor collects data in four visible/near-infrared bands and one panchromatic band. The multispectral bands (ranging from 450 to 900 nm) have spatial resolutions of 2.4 m, while the panchromatic band (450–890 nm) has a 0.6 m pixel resolution. The WorldView-2 sensor collects multispectral data in eight visible and near-infrared channels ranging from 450 to 1,040 nm (2.0 m pixel spatial resolution) and one panchromatic channel (450–800 nm; 0.5 m spatial resolution). In addition to the blue (450–501 nm), green (510–580 nm), red (630–690 nm), and near-infrared (770–895 nm) bands found in QuickBird, four new bands were added to aid in vegetation, soil, and water discrimination—coastal blue (400–450 nm), yellow (585–625 nm), red-edge (705–745 nm), and a second near-infrared (860–1,040 nm) channel.

The QuickBird multispectral data were orthorectified using ground control points (GCPs) obtained in the field. Root mean square (RMS) error for the 2004 image was 0.32 m using 13 field GCPs. The WorldView-2 data were co-registered to the corrected QuickBird image. RMS error of the WorldView-2 image was less than 1 pixel (0.91 m) with 48 GCPs. Following the same methodology, the QuickBird and WorldView-2 panchromatic bands were also co-registered to the rectified multispectral bands using 27 GCPs, with RMS errors of less than one-half pixel.

To make the images compatible for change detection, the WorldView-2 multispectral data were resampled to a 2.4 m×2.4 m pixel size using cubic convolution resampling. The image data were not corrected for atmospheric or radiometric errors due to the lack of available atmospheric parameters at the time of image acquisition over the study area. Clouds and cloud shadows were masked prior to image classification to minimize spectral confusion.

The addition of band ratios, indices, and texture measures has been shown to improve land use/cover classification results (Huang et al. 2002). The simple ratio vegetation index (NIR band/red band) was calculated from the multispectral data, and mean texture was derived from the panchromatic band using a gray-level co-occurrence matrix (GLCM) for each image. An image layer stack consisting of the multispectral bands, vegetation index, and texture measure was created for each image and used as the classification input.

## Field Data and Classification Scheme

In situ land use/cover data were collected in the study area from July to August 2008 and July to August 2009 to provide training and validation data for the classifications.

**Table 9.1** Characteristics of land cover classes identified in the highlands of southern Isabela

| Land use/cover | Description |
| --- | --- |
| Barren | Non-vegetated areas such as exposed soil and lava rock outcrops |
| Built-up | Man-made features including buildings, roads, and structures for animals |
| Crops/pasture/grass | Agricultural areas for crop cultivation, managed pastures, and natural grassland |
| Dry pasture/grass | Dry or senescent vegetation including managed pastures and natural grassland |
| Forest/shrub | Areas covered with dense growth of mostly evergreen trees or taller shrubs, including native and introduced species |
| Guava | Sites dominated by guava (*Psidium guajava*), an invasive woody shrub |
| Rose apple | Areas dominated by dense growth of rose apple (*Syzygium jambos*), an invasive tree |

Sampling areas ($n = 263$) were stratified by land cover type and purposefully selected to capture features of interest, such as patches of invasive species, crops, and buildings. At each location, the land cover type was noted, a site description was recorded, and digital photographs were taken. The observations were geo-located with differentially corrected (post-processing) Global Positioning System (GPS) coordinates. One-third of the sample points ($n = 86$) were used to train the classifications, while the remaining two-thirds ($n = 177$) were reserved for validation.

Seven classes representing the most common land uses and covers in the study area were identified during field visits and selected for image classification: barren, built-up, crops/pasture/grass, dry pasture/grass, forest/shrub, guava, and rose apple (Table 9.1). Guava (*Psidium guajava* L.) and rose apple (*Syzygium jambos* L.) are considered among the worst invaders in the Galápagos Islands because of their ability to significantly transform terrestrial ecosystems (Tye et al. 2002).

## *Object-Based Classification*

Supervised classification of the 2004 and 2010 images was performed with the object-based image analysis approach (OBIA). OBIA is a knowledge-based classification method that attempts to mimic the way humans interpret remote sensing images (Hay and Castilla 2008). Homogenous groups of pixels, or objects, are the basic unit of analysis and thus avoid the "salt-and-pepper" effect in pixel-based classifications of high spatial resolution data (Blaschke et al. 2000). Further, OBIA can exploit the textural, spatial, and topological characteristics of image objects (Lang 2008) to improve the value and accuracy of classifications (Benz et al. 2004). Walsh et al. (2008) successfully mapped guava cover in Isabela's highlands using an OBIA classifier with high spatial resolution satellite data.

**Table 9.2** Segmentation parameters for OBIA classification

| | Input layers | Scale | Color/shape | Compactness/ smoothness |
|---|---|---|---|---|
| *QuickBird image (2004)* | | | | |
| Level 1 | Multispectral bands (4) Simple ratio GLCM texture | 18 | 0.6/0.4 | 0.2/0.8 |
| Level 2 | Multispectral bands (4) GLCM texture | 40 | 0.7/0.3 | 0.2/0.8 |
| *WorldView-2 image (2010)* | | | | |
| Level 1 | Multispectral bands (8) Simple ratio GLCM texture | 18 | 0.6/0.4 | 0.2/0.8 |
| Level 2 | Multispectral bands (8) GLCM texture | 40 | 0.7/0.3 | 0.2/0.8 |

The WorldView-2 data were first segmented into objects with the multiresolution segmentation algorithm in Definiens Professional 5 (Definiens AG, München, Germany). Multiresolution segmentation is a bottom-up, region-merging procedure (Benz et al. 2004) that creates objects corresponding to features of interest in the image without extensive processing times. The goal is to minimize the heterogeneity of extracted image objects while maximizing contrast to neighboring objects. In this study, image objects were generated at two levels through a bottom-up approach. Small objects were created to represent buildings, roads, and other small features (level 1), and a set of larger objects (level 2) were produced to represent vegetation patches, including forests and open fields (Table 9.2). All layers in the image stack were weighted equally, and user-defined criteria describing the threshold for object heterogeneity—scale, color/shape, and smoothness/compactness—were selected iteratively through a visual assessment of object fit (Meinel and Neubert 2004).

The image objects were then classified using a rule-based classification approach. In Definiens Professional, each land use/cover category in the classification scheme contains a set of expressions, or rules, that describe the class. Knowledge-based rules can draw on spectral data contained in the image bands and/or contextual information such as the textural, spatial, and topological characteristics of image objects. Objects corresponding to points in the training data set were isolated, and their spectral, textural, and contextual attributes were used to establish the rules for each class.

The classification algorithm then evaluated the membership value of each image object to the list of classes, and the class with the highest membership value (ranging from 0 to 1) was assigned to the image object. The objects were first separated into "vegetation" and "non-vegetation" classes based on mean simple ratio (SR) vegetation index values. Objects with SR values between 4.5 and 18 were assigned membership in "vegetation," and objects with low membership to the class were categorized as "non-vegetation." "Non-vegetation" objects were further refined into several subclasses (i.e., buildings, lava, dry pasture/grass, and soil) at level 1, while "vegetation" subclasses were defined at level 2 (Table 9.3). The classifications at levels 1 and 2 were then merged to create a single thematic land use/cover map.

**Table 9.3**  QuickBird image (2004): OBIA classification rules including features and membership thresholds

| Final class | Subclasses | Feature | Function[a] and threshold |
|---|---|---|---|
| Barren | Lava | Brightness | <260 |
| | | Mean GLCM texture | 1.35 ⎰9 |
| | | NDVI | <0.3 |
| | Soil | Mean simple ratio | 1 \ 1.5 |
| | | Brightness | 240 ∫ 400 |
| | | NDVI | 0.1 ⎰0.5 |
| Built-up | Building | Area | <306 m² |
| | | Length | <36 m² |
| | | Max difference (to neighbors) | 0 ⎰1.25 |
| | | Mean red band | 190 /-\ 2,250 |
| | Road | Classified as lava or soil | 3–21 |
| | | Length/width | |
| Crops/pasture/grass | Grass—bright | Brightness | 275 ∫ 400 |
| | | Mean red band | 89 ∫ 200 |
| | | NDVI | 0.42–0.67 |
| | Grass—dark | Brightness | 345 ∫ 460 |
| | | Mean green band | 229 ∫ 300 |
| | | Mean red band | 70 ∫ 110 |
| | | NDVI | 0.42–0.67 |
| | Crops/pasture | Brightness | 425 ∫ 460 |
| | | Mean green band | 240 /-\ 330 |
| | | Mean red band | 80 ∫ 110 |
| | | NDVI | 0.5–0.766 |
| Dry pasture/grass | | Brightness | 289 ∫ 370 |
| | | NDVI | 0.1–0.8 |
| Forest/shrub | Trees—green | Mean green band | 260 /-\ 360 |
| | | NDVI | 0.7–0.78 |
| | Trees—yellow | Brightness | 270 ∫ 380 |
| | | Mean green band | 255 ∫ 375 |
| | | NDVI | 0.5–0.7 |
| | | Distance to right image border | 2,500–4,700 m |
| | | Distance to bottom image border | 1,750–3,650 m |
| Guava | | Brightness | 250 ⎰360 |
| | | Mean green band | 230 ⎰300 |
| | | Mean NIR band | 480 ⎰825 |
| | | NDVI | 0.53–0.72 |
| Rose apple | | Brightness | 250 ∫ 340 |
| | | Mean green band | 225 ⎰310 |
| | | NDVI | 0.6–0.74 |
| | | Distance to right image border | 3,850–6,900 m |
| | | Distance to bottom image border | 1,500–5,500 m |

[a]Fuzzy membership functions: ⎰= lower than (nonlinear), ∫= greater than (nonlinear), \= lower than (linear), /= greater than (linear), /-\ = approximate range

The same object-based segmentation and classification approach was applied to the QuickBird image by adjusting the input parameters and threshold values. Image objects at levels 1 and 2 were derived from the image data according to the segmentation parameters in Table 9.2. Training data corresponding areas of invariant land cover (e.g., stable guava patches, established roads) were used to define the membership rules for each class. Objects with SR values between 1.5 and 8.1 were classified as "vegetation," while all other objects were assigned to the "non-vegetation" category. The objects were further classified at levels 1 and 2 based on the classification scheme rules (Table 9.4) and merged into a single output classification, as with the WorldView-2 image.

Accuracy of the 2010 (WorldView-2) classification was assessed with field reference points ($n = 177$) not used as training data during image classification. Standard error matrices were calculated to determine the overall accuracy, producer's and user's accuracies, and overall kappa statistic on a per-pixel basis. Field data to test the accuracy of the 2004 (QuickBird) classification were not available. Post-classification LULC change analysis was performed by overlaying the classified images from 2004 and 2010 and calculating "from–to" change at the pixel level. Change statistics were also generated for the two management zones, Santo Tomás and the Galápagos National Park.

## Sociodemographic Data and Analysis

Data from publicly available secondary data sets and information from interviews with local residents were leveraged to contextualize land use/cover change in Isabela's highlands. The socioeconomic, demographic, and agricultural production factors that likely influence household land use decisions were considered. The secondary data used in this study—Population and Housing Census (2001 and 2010), National Agricultural Census III (2000), and the Galápagos Living Standards Survey (2009)—are publicly available data sets collected and published by the Ecuadorian census agency (INEC).

Demographic changes in Santo Tomás were drawn from the population and housing censuses conducted in 2001 and 2010. Descriptive statistics on the size and age distribution of the population, number and size of households, and primary occupations were calculated in SPSS Statistics v.19 (IBM SPSS Statistics, Chicago, Il.) from individual- and household-level data spatially located at the community level. Information on agricultural production was taken from the agricultural census conducted in 2000 and the 2009 living standards survey. The number and proportion of absentee landowners, products cultivated and quantities harvested, number and types of livestock produced, and the number of farms with hired labor were described from basic statistics generated from household-level data for the entire community of Santo Tomás .

The secondary demographic and agricultural data were supplemented by household interviews conducted with Santo Tomás landholders during July and August 2008. A questionnaire with structured and open-ended questions was administered

**Table 9.4**  WorldView-2 image (2010): OBIA classification rules including features and membership thresholds

| Final class | Subclasses | Feature | Function[a] and threshold |
|---|---|---|---|
| Barren | Lava | Brightness | <280 |
| | | Mean GLCM texture | 1.35⎵9 |
| | | Mean red-edge band | 124⎵375 |
| | Soil | Brightness | 290 ∫445 |
| | | NDVI | 0.2⎵0.6 |
| Built-up | Building | Area | <306 m² |
| | | Length | <36 m |
| | | Max difference (to neighbors) | 0⎵1.75 |
| | | Mean red band | 100 /-\ 2,000 |
| | Road | Classified as lava or soil | |
| | | Length/width | 3–21 |
| Crops/pasture/ grass | Grass—bright | Brightness | 310 ∫420 |
| | | Mean red band | 65 ∫180 |
| | | NDVI | 0.5–0.7 |
| | Grass—dark | Brightness | 355 ∫460 |
| | | Mean green band | 229 ∫300 |
| | | Mean red band | 70 ∫110 |
| | | NDVI | 0.7–0.76 |
| | Crops/pasture | Brightness | 425 ∫460 |
| | | Mean green band | 240 /-\ 330 |
| | | Mean red band | 80 ∫110 |
| | | NDVI | 0.5–0.766 |
| Dry pasture/ grass | | Brightness | 335 ∫405 |
| | | Mean red-edge band | >473 |
| | | NDVI | 0.26 / 0.6 |
| Forest/shrub | Trees—green | Mean green band | 233 /-\ 300 |
| | | NDVI | 0.766–0.84 |
| | Trees—yellow | Brightness | 275 ∫455 |
| | | Mean green band | 230 ∫300 |
| | | NDVI | 0.54–0.735 |
| | | Distance to right image border | 2,500–4,700 m |
| | | Distance to bottom image border | 1,750–3,650 m |
| Guava | | Brightness | 270⎵425 |
| | | Mean green band | 195⎵260 |
| | | Mean NIR-2 band | 540⎵1,045 |
| | | NDVI | 0.61–0.8 |
| Rose apple | | Brightness | 200 ∫427 |
| | | Mean green band | 200⎵240 |
| | | NDVI | 0.7–0.8 |
| | | Distance to right image border | 3,850–6,900 m |
| | | Distance to bottom image border | 1,200–5,500 m |

[a]Fuzzy membership functions: ⎵=lower than (nonlinear), ∫=greater than (nonlinear), \=lower than (linear), /=greater than (linear), /-\=approximate range

to the heads of 45 households and/or their spouses (representing approximately 23% of landholders in Santo Tomás ) using a purposeful sampling scheme.[1] The interviews included questions about household demographics, land use patterns, invasive plants, and changes in the community over the last decade. Patterns in the data were analyzed with particular attention to changes in agricultural land use and invasive plant cover.

# Results

## *Land Use/Cover Classification and Change Detection*

Overall accuracy of the 2010 classification was 88.70%, with a kappa statistic of 0.87 (Table 9.5). Although overall accuracy exceeded the 85% threshold (Foody 2002), forest/shrub cover was not as accurately classified. Forest and shrub patches were confused with guava in areas where taller trees cast shadows on neighboring vegetation and resulted in some forested objects being misclassified as guava because of similar spectral responses. The forest class also suffered from errors of commission, particularly due to the misclassification of agriculture and grassland as forest and shrub. Spectral confusion between these classes may be the result of the spectral heterogeneity of pixels used to train the crops/pasture/grass class. Field data to test the accuracy of the 2004 classification were not available, but the same classification approach was applied to both images in an effort to produce classifications with comparable accuracies. Visual assessment of the 2004 classification showed that invariant features, such as the Sierra Negra caldera, main roads, and surface mines, were correctly classified.

Comparison of the land cover classifications reveals significant land use/cover conversion between 2004 and 2010 (Table 9.6, Fig. 9.2). Across the study area, guava remained the most dominant land cover, increasing from 35.5 to 39.7% of the landscape. The largest expansion of guava occurred in the national park, where an additional 273 ha of land were invaded between 2004 and 2010 (Table 9.7). Santo Tomás experienced only a small net gain in guava (2.2%). However, guava is by far the most dominant land cover in the community and covers nearly 47% of the agricultural zone. The largest patches of stable guava, corresponding to fields and entire farms in some cases, are located in western and northern Santo Tomás . New areas of invasion (since 2004) are smaller and occur adjacent to existing patches within Santo Tomás and to the north and south along the national park border.

Crops/pasture/grass occupied an extensive area in 2004 (28.8%) that declined to just over 20% of the landscape in 2010 (Table 9.6). Agriculture in Santo Tomás

---

[1] A random sampling strategy was originally intended but had to be adapted after it was revealed that cadastral maps used to locate properties were more than 30 years old and no longer accurate. The small number of households still living in Santo Tomás as well as landholders now residing in Puerto Villamil were interviewed to approximate planned sampling levels.

**Table 9.5** Confusion matrix for 2010 WorldView-2 classification

| Mapped class | Reference class | | | | | | | | |
| | Barren | Built-up | Crops/ pasture/ grass | Dry pasture/ grass | Forest/ shrub | Guava | Rose apple | Total | User's accuracy |
|---|---|---|---|---|---|---|---|---|---|
| Barren | 31 | 0 | 0 | 1 | 0 | 0 | 0 | 32 | 96.9% |
| Built-up | 1 | 33 | 0 | 0 | 0 | 0 | 0 | 34 | 97.1% |
| Crops/pasture/ grass | 0 | 0 | 26 | 0 | 3 | 0 | 0 | 29 | 89.7% |
| Dry pasture/grass | 0 | 0 | 1 | 13 | 0 | 2 | 0 | 16 | 81.2% |
| Forest/shrub | 0 | 1 | 3 | 1 | 20 | 1 | 1 | 27 | 74.1% |
| Guava | 0 | 0 | 3 | 0 | 2 | 23 | 0 | 28 | 82.1% |
| Rose apple | 0 | 0 | 0 | 0 | 0 | 0 | 11 | 11 | 100.0% |
| Total | 32 | 34 | 33 | 15 | 25 | 26 | 12 | 177 | – |
| Producer's accuracy | 96.9% | 97.1% | 78.8% | 86.7% | 80.0% | 88.5% | 91.7% | – | – |

Overall = 88.70%
Kappa = 0.87

**Table 9.6** Land use/cover area and change (net area, percent relative to 2004), 2004–2010, Isabela highlands

| Land use class | Total area (ha) | | Percent of landscape (%) | | Change: 2004–2010 | |
| | 2004 | 2010 | 2004 | 2010 | Absolute (ha)[a] | Relative (%)[b] |
|---|---|---|---|---|---|---|
| Barren | 208.10 | 467.75 | 2.8 | 6.2 | 259.65 | 124.8 |
| Built-up | 23.65 | 26.11 | 0.3 | 0.3 | 2.45 | 10.4 |
| Crops/pasture/ grass | 2,167.76 | 1,569.64 | 28.8 | 20.8 | −598.12 | −27.6 |
| Dry pasture/ grass | 418.53 | 278.63 | 5.6 | 3.7 | −139.91 | −33.4 |
| Forest/shrub | 1,992.51 | 2,119.34 | 26.4 | 28.1 | 126.83 | 6.4 |
| Guava | 2,673.34 | 2,992.09 | 35.5 | 39.7 | 318.75 | 11.9 |
| Rose apple | 49.71 | 80.06 | 0.7 | 1.1 | 30.35 | 61.1 |
| Total | 7,533.61 | 7,533.61 | 100.0 | 100.0 | – | – |

[a]Net change between periods was calculated as (Area2010 − Area2004)
[b]Percent change relative to 2004 was calculated as 100 × (Area2010 − Area2004)/Area2004

declined by 28.8% (relative to 2004) (Table 9.7). A few, small patches of land were brought into agricultural production between 2004 and 2010 (totaling 389 ha), primarily in northern and eastern Santo Tomás . However, more than 800 ha of land in crops/pasture/grass were converted to other land covers like guava, dry pasture (a less intensive agricultural use), and forest. In the national park, where agricultural land use is prohibited, grasslands were transformed to guava along the caldera and to forest/shrub in the transition zone to the east (Fig. 9.2).

Although forest cover experienced a net increase across the study site, from 26.4% of the landscape in 2004 to 28.1% in 2010, opposing trends were observed in the national park and Santo Tomás (Table 9.6). Forest/shrub cover in the national park remained

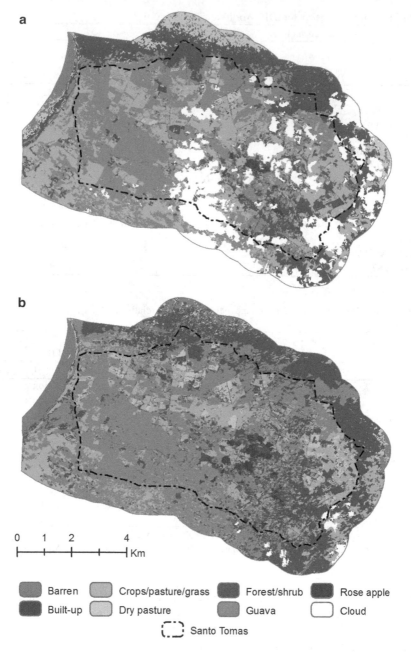

**Fig. 9.2** Land use/cover in the study area in 2004 (**a**) and 2010 (**b**)

**Table 9.7** Net change in land cover from 2004 to 2010 as a proportion of each management zone

| Land use class | Area of management zone (ha) | | Percent of management zone (%) | | Change: 2004–2010 | |
|---|---|---|---|---|---|---|
| | 2004 | 2010 | 2004 | 2010 | Absolute (ha)[a] | Relative (%)[b] |
| *Santo Tomás* | | | | | | |
| Barren | 48.49 | 87.75 | 1.1 | 1.9 | 39.26 | 81.0 |
| Built-up | 21.00 | 23.91 | 0.5 | 0.5 | 2.91 | 13.9 |
| Crops/pasture/grass | 1,449.45 | 1,032.70 | 32.2 | 22.9 | −416.75 | −28.8 |
| Dry pasture/grass | 82.35 | 216.43 | 1.8 | 4.8 | 134.08 | 162.8 |
| Forest/shrub | 798.01 | 963.62 | 17.7 | 21.4 | 165.61 | 20.8 |
| Guava | 2,058.99 | 2,104.27 | 45.7 | 46.7 | 45.28 | 2.2 |
| Rose apple | 49.70 | 79.29 | 1.1 | 1.8 | 29.59 | 59.5 |
| Total | 4,507.97 | 4,507.97 | 1.1 | 1.9 | – | – |
| *Galápagos National Park* | | | | | | |
| Barren | 159.62 | 380.00 | 5.3 | 12.6 | 220.38 | 138.1 |
| Built-up | 2.65 | 2.20 | 0.1 | 0.1 | −0.45 | −17.2 |
| Crops/pasture/grass | 718.31 | 536.94 | 23.7 | 17.7 | −181.37 | −25.2 |
| Dry pasture/grass | 336.19 | 62.19 | 11.1 | 2.1 | −274.00 | −81.5 |
| Forest/shrub | 1,194.51 | 1,155.72 | 39.5 | 38.2 | −38.79 | −3.2 |
| Guava | 614.35 | 887.82 | 20.3 | 29.3 | 273.47 | 44.5 |
| Rose apple | 0.01 | 0.78 | 0.0 | 0.0 | 0.77 | 9,514.3 |
| Total | 3,025.64 | 3,025.64 | 5.3 | 12.6 | – | – |

[a]Net change between periods was calculated as (Area2010−Area2004)
[b]Percent change relative to 2004 was calculated as 100×(Area2010−Area2004)/Area2004

largely unchanged, declining by only 3.2%. In Santo Tomás , forest/shrub increased as a result of conversion of agriculture and guava, as previously mentioned.

The increase in barren land since 2004 (124.8%) resulted from new lava rock that covered the caldera of Sierra Negra following its eruption in 2005, an area in the north that transitions between dry vegetation and bare soil, and small clearings in Santo Tomás . Built features did not change substantially between 2004 and 2010, making up only 0.3% of the landscape (0.5% of Santo Tomás ) (Table 9.7). Rose apple, which also made up a small percentage of the total landscape in 2004, spread within central Santo Tomás . The area of invasion increased from 49.7 ha in 2004 (1.1%) to 79.29 ha in 2010 (1.8%). Although rose apple was restricted to Santo Tomás in 2004, by 2010, it had expanded into the national park, covering 0.78 ha of land.

## Sociodemographic Trends

The census data reveal interesting population shifts in Santo Tomás . Between 2001 and 2010, total population declined by 17.6%, at a rate of 2.2% per annum (Table 9.8). The number of households in Santo Tomás also declined, while mean

**Table 9.8** Demographic indicators and agricultural production for Santo Tomás , 2000–2009

|  | 2001 | 2010 |
| --- | --- | --- |
| Population (total)[a] | 199 | 164 |
| Number of households | 66 | 54 |
| Household size (mean) | 2.97 | 3.04 |
| Age (median) | 27 | 32 |
|  | 2000 | 2009 |
| Landholders living in Santo Tomás  (%) | 40.7 | 22.3 |
| Farms cultivating annuals/perennials (%) | 81.5 | 80.8 |
| Harvest sold (%) | 61.7 | 21.1 |
| Cattle | 1,972 | 888 |
| Hogs | 236 | 105 |
| Farms with paid laborers (%) | 37 | 25.4 |

[a]Includes floating (tourist) population

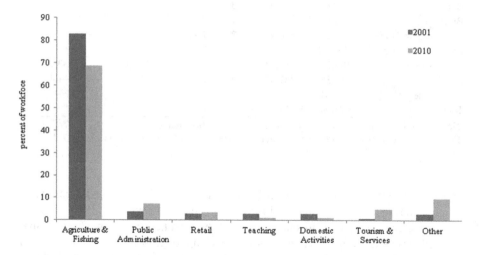

**Fig. 9.3** Proportion of Santo Tomás  workforce employed in various sectors in 2001 and 2010

household size was relatively unchanged. A total of 66 households resided in Santo Tomás  with an average of 2.97 members in 2001. By 2010, only 54 households remained. Median age for Santo Tomás  increased slightly, from 27 years in 2001 to 32 years in 2010. Agriculture and fishing remained the largest employment sectors in Santo Tomás , despite significant increases in other categories (Fig. 9.3). While 83% of working age residents (15–64) reported agriculture or fishing as their primary occupation in 2001, only 68% participated in the sector by 2010.

The agricultural census and living standards survey provide additional details about the state of agricultural production and farming households. Between 2000 and 2009, the proportion of landholders who still lived on the farms declined from 40.7% to just over 22% (Table 9.8). Although the proportion of farms cultivating annual and/or perennial crops decreased only slightly, the majority of crops harvested in 2009 were not sold. With respect to livestock production, there was a net reduction in the number of cattle and hogs raised in Santo Tomás . Finally, in 2009, fewer farms hired laborers (25.4%) to assist with agricultural activities like clearing and planting than in 2000 (37%).

## Discussion

The rural community of Santo Tomás and adjacent land managed by the Galápagos National Park experienced substantial land use/cover changes from 2004 to 2010. The change detection analysis revealed a substantial decline (nearly 29%) in agricultural land use observed in 2010 compared with 2004 (Table 9.6). While some new areas were brought into production between 2004 and 2010, a significant amount of land (800 ha) was converted to less productive pastures (dry pasture) or transformed to woody vegetation including guava and forest/shrub (Fig. 9.2). Production data (Table 9.8) show that as agricultural land use has declined, production has also become less intense. The majority of annual and perennials grown in Santo Tomás in 2009 were not sold, and fewer livestock were reared than in the earlier period. Further, off-farm employment opportunities have increased, and fewer working age adults in Santo Tomás participate in the agricultural sector (Fig. 9.3).

These results seem to suggest that over the past decade, many households have abandoned agriculture, choosing instead to participate in off-farm activities to support the household (Table 9.8). During interviews, many heads of household noted a lack of diversity in what farms produce. Further, a market for their products does not exist on Isabela or other islands, limiting the income that can be derived from agriculture. Isabela's most recent strategic plan noted that agricultural production is not sufficient to reliably satisfy local demand throughout the year, so fruits, vegetables, and dairy products have to be imported from continental Ecuador (Vilema et al. 2003). In interviews, landholders also described a variety of barriers to farming, including the lack of freshwater for household use and irrigation, the presence of various pests, the lack of financing (i.e., access to credit), and limited technical assistance.

The abandonment of agricultural activities appears to be coupled with rural emigration and abandonment of land in the highlands. Population decline in Santo Tomás (Table 9.8) and an increase in the proportion of the population residing in the urban community (nearly 93%) likely reflect outmigration from rural areas. Interview data suggest that most landholders live in Puerto Villamil, the urban community south of Santo Tomás , and visit their farms only occasionally. Agriculture on Isabela is not mechanized, and rather than hiring additional laborers to maintain productivity (Table 9.8), many farms allow land to lie fallow indefinitely. The availability of employment opportunities in tourism and the service industry catering to tourists (Fig. 9.3) may be another factor driving emigration (Kerr et al. 2004). Isabela is not unique in this respect, as urban–rural migration and farm abandonment have been observed elsewhere in Galápagos (Rodriguez 1989; Kerr et al. 2004; Borja and Perez 2000).

The highlands have experienced significant increases in guava and forest/shrub cover at the expense of agricultural land (Table 9.6). According to interviews, farmers are no longer purposefully cultivating guava. Rather, it has become naturalized and now grows unaided throughout the highlands. Guava and other introduced plants, like rose apple, can spread rapidly in abandoned lands, directly contributing to the expansion of invasive species into the national park (Borja and Perez 2000; Walsh et al. 2008). Discussions with households demonstrate that farmers recognize the important of clearing guava, but doing so is time consuming and expensive. Due to the cost of manual removal and the need for control measures at regular intervals (every 6 months), some owners have chosen to abandon lands that are seriously invaded. The land cover analysis also demonstrated increasing forest/shrub cover in the last decade. Encroachment of introduced and invasive trees into formerly treeless vegetation zones in the highlands (Miconia and fern–sedge communities, sensu Wiggins and Porter 1971) may alter local environmental conditions and lead to declines in species diversity and native/endemic plant cover (Jäger et al. 2009).

## Conclusions

This chapter provides an enhanced understanding of contemporary land use/cover in the highlands of southern Isabela Island and points to a few of the processes driving land cover conversion—land abandonment, declining agricultural production, and the spread of invasive plants. The findings presented here are consistent with those reported by Villa and Segarra (2010) who found that agricultural land on San Cristobal Island was abandoned between 1987 and 2000 due to low returns on production and labor constraints. Future studies should attempt to quantify the socio-economic and environmental factors that drive patterns of land use/cover change and landscape dynamics on Isabela Island. Empirical data on the impacts of changing land use/cover on biodiversity and ecosystem functioning in the highlands is limited and warrants attention.

In addition, this study offers a methodological approach to the assessment of land use/cover change that could be applied elsewhere in the Galápagos. Remote sensing provides an effective method for mapping spatial patterns of land use/cover and for quantifying spatial patterns and rates of change. The description of land use change and its driving forces can provide important information for land managers and decision makers in the archipelago. Several applications in Galápagos have been recognized, ranging from the generation of more complete information on species distributions (Trueman et al. 2010) and the development of weed risk assessment systems (Tye et al. 2002) to regional planning of natural resources (Villa and Segarra 2010) and identifying barriers to conservation and restoration projects (Gardener et al. 2010).

**Acknowledgments**  This work was funded in part by an NSF IGERT grant (DGE-0333193) to the Carolina Population Center at the University of North Carolina at Chapel Hill. Support for this research was provided by the Center for Galápagos Studies and the Department of Geography at the University of North Carolina at Chapel Hill, the Galápagos National Park Service, and the people of Isabela Island. The comments of an anonymous reviewer are also appreciated.

# References

Alo CA, Pontius RG Jr (2008) Identifying systematic land-cover transitions using remote sensing and GIS: the fate of forests inside and outside protected areas of Southwestern Ghana. Environ Plann B 35:280–295

Benz UC, Hofmann P, Willhauck G, Lingenfelder I, Heynen M (2004) Multi-resolution, object-oriented fuzzy analysis of remote sensing data for GIS-ready information. ISPRS J Photogramm 58:239–258

Blaschke T, Lang S, Lorup E, Stobl J, Zeil P (2000) Object-oriented image processing in an integrated GIS/remote sensing environment and perspectives for environmental applications. In: Cremers A, Greve K (eds) Environmental information for planning, politics and the public. Verlag, Marburg

Borja RN, Perez J (2000) Parque nacional Galápagos. dinamicas migratorias y sus efectos en el uso de la recursos naturales. Fundacion Natura and The Nature Conservancy, Quito

Brandt JJS, Townsend PA (2006) Land use—land cover conversion, regeneration and degradation in the high elevation Bolivian Andes. Landsc Ecol 21:607–623

Cléder E, Grenier C (2010) Taxis in Santa Cruz: uncontrolled mobilization. Galápagos report 2009–2010. CDF, GNP, CGG, Puerto Ayora, Galápagos

Collins A, Bush MB (2011) An analysis of modern pollen representation and climatic conditions on the Galápagos Islands. Holocene 21:237–250

Foody GM (2002) Status of land cover classification accuracy assessment. Remote Sens Environ 80:185–201

Froyd C, Lee J, Anderson A, Haberle S, Gasson P, Willis K (2010) Historic fuel wood use in the Galápagos Islands: identification of charred remains. Veg Hist Archaeobot 19:207–217

Gardener MR, Grenier C (2011) Linking livelihoods and conservation: challenges facing the Galápagos Islands. In: Baldacchino G, Niles D (eds) Island futures: conservation and development across the Asia-Pacific region. Springer, New York

Gardener MR, Atkinson R, Rentería JL (2010) Eradications and people: lessons from the plant eradication program in Galápagos. Restor Ecol 18:20–29

Gonzalez JA, Montes C, Rodriguez J, Tapia W (2008) Rethinking the Galápagos Islands as a complex social–ecological system: implications for conservation and management. Ecol Soc 13(2):13

Guézou A, Trueman M, Buddenhagen CE, Chamorro S, Guerrero AM, Pozo P, Atkinson R (2010) An extensive alien plant inventory from the inhabited areas of Galápagos. PLoS One 5(4):e10276

Hay GJ, Castilla G (2008) Geographic object-based image analysis (GEOBIA): a new name for a new discipline. In: Blaschke T, Lang S, Hay GJ (eds) Object-based image analysis: spatial concepts for knowledge-driven remote sensing applications. Springer, Berlin

Henderson S, Dawson TP (2009) Alien invasions from space observations: detecting feral goat impacts on Isla Isabela. Galápagos Islands with the AVHRR. Int J Remote Sens 30:423–433

Huang C, Davis LS, Townshend JRG (2002) An assessment of support vector machines for land cover classification. Int J Remote Sens 23:725–749

Jäger H, Kowarik I, Tye A (2009) Destruction without extinction: long-term impacts of an invasive tree species on Galápagos highland vegetation. J Ecol 97:1252–1263

Kerr S, Cardenas S, Hendy J (2004) Migration and the environment in the Galápagos: an analysis of economic and policy incentives driving migration, potential impacts from migration control, and potential policies to reduce migration pressure. Motu Economic and Public Policy Research, Wellington

Lang S (2008) Object-based image analysis for remote sensing applications: modeling reality—dealing with complexity. In: Blaschke T, Lang S, Hay GJ (eds) Object-based image analysis: spatial concepts for knowledge-driven remote sensing applications. Springer, Berlin

MacFarland C, Cifuentes M (1996) Case study: Galápagos, Ecuador. In: Dompka V (ed) Human population, biodiversity and protected areas: science and policy issues. Washington, DC, American Association for the Advancement of Science

Meinel G, Neubert M (2004) A comparison of segmentation programs for high resolution remote sensing data. Int Arch ISPRS 35:1097–1105

Ministerio de Agricultura y Ganaderia Programa Nacional de Regionalizacion Agraria (PRONAREG), Institut Francais de Recherche Scientifique Pour le Developpement en Cooperation (ORSTOM), Instituto Nacional Galápagos (INGALA) (1987) Islas Galápagos: Mapa de Formaciones Vegetales (1:100000)

Rodriguez J (1989) Una agricultura exigua en un espacio rural singular. Trama 49:13–15

The Nature Conservancy (TNC), Centro de Levantamientos Integrados de Recursos Naturales por Sensores Remotos (CLIRSEN) (2006) Cartografía Galápagos 2006: conservación en otra dimensión (1:50,000)

Trueman M, Hannah L, d'Ozouville N (2010) Terrestrial ecosystems in Galápagos: potential responses to climate change. In: Ona IL, Di Carlo G (eds) Climate change vulnerability assessment of the Galápagos Islands. Quito, World Wildlife Fund and Conservation International

Tye A, Soria M, Gardener M (2002) A strategy for Galápagos weeds. In: Veitch CR, Clout MN (eds) Turning the tide: the eradication of invasive species (proceedings of the International Conference of Eradication of Island Invasives). Aukland, New Zealand, IUCN Species Survival Commission

Vilema H, Carrion C, Gordillo P (2003) Plan Estrategico del Canton Isabela. Gobierno Municipal de Isabela, Puerto Villamil

Villa A, Segarra P (2010) El cambio histórico del uso del suelo y cobertura vegetal en el área rural de Santa Cruz y San Cristóbal. Galápagos report 2009–2010. CDF, GNP, CGG, Puerto Ayora, Galápagos

Walsh SJ, McCleary AL, Mena CF, Shao Y, Tuttle JP, González A, Atkinson R (2008) QuickBird and Hyperion data analysis of an invasive plant species in the Galápagos Islands of Ecuador: implications for control and land use management. Remote Sens Environ 112:1927–1941

Walsh SJ, McCleary AL, Heumann BW, Brewington L, Raczkowski EJ, Mena CF (2010) Community expansion and infrastructure development: implications for human health and environmental quality in the Galápagos Islands of Ecuador. J Lat Am Geogr 9:137–159

Watson J, Trueman M, Tufet M, Henderson S, Atkinson R (2009) Mapping terrestrial anthropogenic degradation on the inhabited islands of the Galápagos Archipelago. Oryx 44:79–82

Wiggins IL, Porter DM (1971) Flora of the Galápagos Islands. Stanford University Press, Stanford, CA

# Chapter 10
# Investigating the Coastal Water Quality of the Galapagos Islands, Ecuador

Curtis H. Stumpf, Raul A. Gonzalez, and Rachel T. Noble

## Introduction

The Galapagos Archipelago is a UNESCO World Heritage Site and unique ecological setting of species diversity. Internationally renowned for its link to Charles Darwin's research and seminal publication on biodiversity, "Origin of the Species," a lesser known fact is that the archipelago is home to more than 30,000 inhabitants, a rapidly growing population, and a burgeoning tourism industry (Epler 2007; Watkins and Cruz 2007). Tourism has grown dramatically in recent years, increasing more than threefold from 1990 to 2006 (Watkins and Cruz 2007), with more than 170,000 people visiting the islands in 2010 (GNPS 2011). The increase in tourism (and economic growth) has resulted in increased immigration (Kerr et al. 2004). This interconnected growth in tourism and population has created a commensurate if not greater strain on the limited infrastructure and ecological resource of the islands (Walsh et al. 2010). Unsustainable growth threatens the islands' resources, the tourism-driven economy, and human health through inadequate infrastructure, especially pertaining to water resources.

Water resources are critically important to the Galapagos Islands (Hennessy and McCleary 2011; Lopez and Rueda 2010; Kerr et al. 2004). Management of marine and fresh water is paramount to the success and balance of the Galapagos Island economy and ecology; however, the human impact on groundwater and marine water resources is apparent in bacterially contaminated aquifers on Santa

C.H. Stumpf (✉)
Crystal Diagnostics Ltd., 4209 State Route 44, Rootstown,
OH 44272, USA
e-mail: cstumpf@crystaldiagnostics.com

R.A. Gonzalez • R.T. Noble
Institute of Marine Sciences, University of North Carolina at Chapel Hill,
3431 Arendell Street, Morehead City, NC 28557, USA
e-mail: rgonzale@live.unc.edu; rtnoble@email.unc.edu

S.J. Walsh and C.F. Mena (eds.), *Science and Conservation in the Galapagos Islands:*
*Frameworks & Perspectives*, Social and Ecological Interactions in the Galapagos Islands 1,
DOI 10.1007/978-1-4614-5794-7_10, © Springer Science+Business Media, LLC 2013

Cruz Island, contaminated household water on San Cristobal Island, and contamination of nearshore marine waters on Isabela Island (d'Ozouville 2008; Lopez and Rueda 2010; Walsh et al. 2010). Wastewater management is one of the greatest challenges, as most population centers depend on septic (onsite wastewater treatment systems usually consisting of a collection area and subsequent leaching into the ground) or poorly constructed sewage systems (Walsh and McCleary 2009), which are not well suited for the lava-like bedrock which contains little soil and primarily fractured basalt (d'Ozouville 2008). This lack of sewage infrastructure and non-ideal subsurface create a high potential for water resource contamination.

Internationally, fecal contamination of coastal waters used for recreation and seafood production is of serious concern given the potential public health risk associated with contact and/or ingestion of fecal pathogens. In developing countries with increasing populations and insufficient infrastructure, fecal contamination can be of concern to both inhabitants and visitors (Rose 2006). Health risks often include gastrointestinal and respiratory illness, skin and eye irritation and, if symptoms go untreated, can result in more severe illness (Fleisher et al. 2010; Wade et al. 2010). Water-related illnesses have been reported in the Galapagos Islands, though are often anecdotal and poorly researched (Hennessy and McCleary 2011; Walsh et al. 2010).

The impact of fecal contamination, such as that from sewage discharge, also extends to an array of animals and plants (Fernandez 2008; Werdeman 2006). Sewage treatment plants often discharge waters that have high concentrations of heavy metals, organic compounds, detergents, endocrine disruptors, and personnel care products that can have myriad effects on wildlife (Brausch and Rand 2011; Islam and Tanaka 2004; Atkinson et al. 2003). Even the most advanced sewage treatment plants discharge highly labile organic and inorganic nutrients, often contributing directly to the formation of surface algae and phytoplankton growth in the water. These surface "blooms" can contribute to low dissolved oxygen in coastal waters (Pearl 2009), endangering the survival of fish, turtles, amphibians, and benthic organisms with consequences that are severe (Fernandez 2008).

## Galapagos Islands Water Resources

### Water Impairment

Fecal microbial contamination of water resources of the Galapagos Islands, though anecdotally suspected, has been difficult to demonstrate scientifically. Water quality studies have mainly focused on freshwater (groundwater and household drinking water) and have found elevated *Escherichia coli* (*E. coli*) and fecal coliform concentrations (Lopez and Rueda 2010). A recent study at eight total sites on three islands, conducted by the Galapagos National Park in combination with the Japan International Cooperation Agency (JICA), found levels of fecal coliforms (of which *E. coli* is a major subset) in groundwater, lagoon water, and household water at levels ranging from $10^2$ to $10^3$ fecal coliform bacteria per 100 ml, for all sample types (Lopez and Rueda 2010). These levels are beyond limits established by Ecuadorian national environmental legislation to protect public health (TULAS 2003). Marine and brackish water fecal contamination has been previously determined around larger towns of the islands and their associated ports (Kerr et al. 2004; Moir and Armijos 2007; d'Ozouville 2008; Lopez and Rueda 2010). For instance, a study by Armijos et al. (2002) in the coastal waters surrounding Puerto Ayora, Santa Cruz, found marine water degradation due to high *E. coli* concentrations at sites where contaminated groundwater leachate was flowing into Academy Bay during low tide, which is when water is being pulled from the land into the coastal environment.

Overall, long-term microbial water quality monitoring reports and robust data sets are not available for the Galapagos Islands (Walsh et al. 2010), especially for nearshore marine waters. When microbial water quality data does exist, it does not indicate sources of contamination, routes of transport, or seasonality. These are all more difficult pieces of information to derive. In addition, all previous reported monitoring has been conducted using non-molecular methods (e.g., multiple tube fermentation or most probable number methods such as direct culture plating or utilization of chromogenic substrate tests). Though these studies of microbial water quality are useful, they are generally insufficient for identifying the source(s) of contamination, which are likely to be variable and multiple. Using a sanitary survey style to visually identify potential sources of contamination, and based in previous research, the following are suspected sources of fecal contamination to marine waters: contaminated groundwater (Lopez and Rueda 2010), submarine sewage discharges, overland pipe discharges (Walsh and McCleary 2009), boat discharges (Werdeman 2006), and surface runoff during wet weather. Source-specific monitoring is needed to understand the complete picture of potential water impairment in the Galapagos Islands.

Elevated nutrient levels and decreased dissolved oxygen concentrations have also been reported within Academy, Wreck, and Turtle Bay waters proximate to some of the largest towns in the Galapagos Islands and are suspected to be related to sewage contamination (Fernandez 2008; Werdeman 2006). A study by Werdeman (2006) examined the marine waters around Puerto Ayora, Puerto Baquerizo Moreno, and Puerto Villamil and compared nutrient levels (phosphate, nitrate, and ammonium) to a non-populated reference bay (Cartago Bay). Generally, levels of nutrients were higher in all populated bays than the reference bay and showed increased eutrophication and decreased dissolved oxygen. A second study conducted by Fernandez (2008) also found increased nutrient levels in Academy Bay, the bay surrounding Puerto Ayora, Santa Cruz. This study determined a net flux of nutrients from the terrestrial to the marine environment, leading to potential increases in both eutrophication and microbial contamination. In both studies, it was concluded that these elevated nutrient concentrations were partially due to sewage inputs (Fernandez 2008; Werdeman 2006).

## Water Infrastructure

Throughout the Galapagos Islands, expanding population and community sprawl are challenging an already stressed and inadequate water and sanitation infrastructure. In a domino effect scenario, increased population and sprawl leads to increased sewage waste, which increases unregulated discharge and overwhelms an already overburdened wastewater system. Without proper infrastructure, "straight piping" into fissures (Moir and Armijos 2007), septic systems in subsurfaces not ideal for this method of disposal (d'Ozouville 2008), and cesspools (latrines) are often utilized (Walsh and McCleary 2009). Fecal contamination is occuring in coastal waters near growing population centers like Puerto Ayora, where brackish lagoons have been contaminated (Kerr et al. 2004; Moir and Armijos 2007), as well as Isabela, where previous untreated waste was directly piped to lagoons and mangroves (Walsh and McCleary 2009), and even on San Cristobal (where greater than 85% of households are connected to a public sewage disposal system) where untreated sewage is pumped through town to a submarine discharge, resulting in increased eutrophication and microbial contamination of waters proximate to Punta Carola, a popular swimming and surfing area. In essence, tourism and population growth in the Galapagos Islands are out of balance with the development and demand for water and wastewater infrastructure (Hennessy and McCleary 2011).

## Impacts

Water quality impairment has human health implications for users of marine waters for swimming. As described earlier, fecal contamination of water resources can have major repercussions for the public health of residents and tourists alike in the Galapagos Islands. Though a paucity of data exists to link water quality contamination directly with public illness (Walsh et al. 2010), anecdotal evidence from tourists and residents of the islands points to direct connections between water contamination and illness (Hennessy and McCleary 2011). For instance, Hennessy and McCleary (2011) reported that, based on information from a physician of Puerto Villamil, Isabela, as many as 70% of local illnesses were related to contact (either through consumption or exposure) to contaminated water. These illnesses reported are typical of fecal microbial water quality contamination, such as gastroenteritis and skin diseases, and commonly affect younger children (Wade et al. 2010; Colford et al. 2007). Generally, waterborne illnesses are underreported, indicating that this correlation between water contact and illness may be even greater than estimated.

Nearshore marine water quality degradation not only threatens the public health of tourists and residents but also impacts the wildlife of the archipelago. Previous research has shown impacts of sewage contamination to mollusk and crabs assemblages (Cannicci et al. 2009), large mammals such as sea lions (Sturm et al. 2011), general fish species through impacts to food sources such as phytoplankton (Pearl 2009), and chemicals associated with altered breeding and reproduction (Al-Bahry et al. 2009; Penha-Lopes et al. 2009). Additional impacts to wildlife have also been reported from increased heavy metals and endocrine-disrupting compounds (Brausch and Rand 2011). Within the Galapagos Islands, marine wildlife includes sally lightfoot crabs, sea lions, marine iguanas, and myriad fish species. Numerous birds, including the blue-footed boobies, the wingless cormorant, and the magnificent frigate bird, are also potentially affected through their dependence on marine-based food sources. These species are likely affected by decreased nearshore water quality, which results in fish kills and effects reproduction and recruitment of many lower trophic food species.

One example of potential sewage impacts on wildlife of the Galapagos Islands is the sea lion. An integral part of the Galapagos Islands marine landscape and a tourism icon, sea lions may be affected by sewage-related marine water impairment. Previous studies have shown that these pinnipeds are particularly susceptible to illness when inbreeding increases and pathogens are common (Sturm et al. 2011; Acevedo-Whitehouse et al. 2003), which is likely the case for sea lions in the Galapagos Islands' marine environment. Even though sea lion populations have recovered (following a massive reduction in recruitment during the 1997–1998 El Niño), current offspring are experiencing a high degree of illness (Jiménez-Uzcátegui et al. 2007). These high rates of illness may have some links to stressors associated with human-caused marine water impairment. For instance, recent research by Sturm et al. (2011) found increased illness in sea lions on the Chilean coast due to *Salmonella enterica* infection and identified exposure to sewage as one of the likely causes. For a place which ecologically thrives on a tourism industry driven by the image of a pristine and healthy ecological landscape, determining whether water contamination is leading to wildlife health impacts and ameliorating such impacts is critical.

## Characterizing Fecal Contamination in the Coastal Waters of Santa Cruz and San Cristobal Using Molecular Methods

### Study Area and Sample Analysis

A small-scale study was conducted on two islands using molecular techniques to determine the quantities of *Enterococcus* spp. and *Bacteroides* spp. specific markers. Samples were collected within the cities of Puerto Baquerizo Moreno, San Cristobal, and Puerto Ayora, Santa Cruz (Fig. 10.1). These population centers were proximal to coastal waters that were considered impaired by visual observation. Other relatively non-impaired sites adjacent to these populated areas were also sampled as reference sites. Samples were filtered and stored frozen for later analysis at the University of North Carolina at Chapel Hill's Institute of Marine Science, a laboratory with access to advanced tools for analysis.

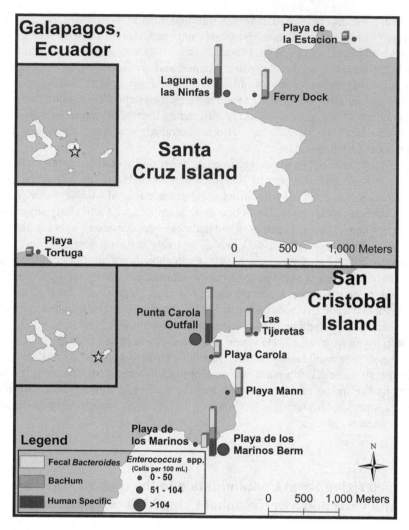

**Fig. 10.1** Study sites sampled on San Cristobal and Santa Cruz Islands, Galapagos Islands

Molecular analysis of samples to quantify *Enterococcus* spp. and source-specific *Bacteroides* spp. markers followed existing published protocols (Noble et al. 2010; Converse et al. 2009; Kildare et al. 2007; Seurinck et al. 2005). The trio of *Bacteroidales* spp. marker assays, fecal *Bacteroides* spp., BacHum, and HF183 (human specific), cover a gradient of specificity and sensitivity. The fecal *Bacteroides* spp. assay is the least specific and quantifies a cohort of anaerobic bacterial species that are found most closely associated with human feces but can also be found in some animal fecal material in lower concentrations. The BacHum and human-specific assays are more specific to human fecal contamination, with the human-specific assay having the greatest ability to discriminate between animal and human fecal contamination (94–100% discriminatory ability, Ahmed et al. 2009). In real-world samples, quantification of both the human-specific and BacHum markers, along with high concentrations of the fecal *Bacteroides* spp. is indicative of a strong potential for the presence of human fecal contamination in a body of water.


## Study Results and Implications


*Enterococcus* spp. concentrations were elevated (greater than 104 CE/100 ml) in 85.7% (6 of 7) of the samples in the coastal waters of Punta Carola Outfall and Playa de los Marinos Berm, San Cristobal. Analysis of the samples using the trio of Bacteroidales-based molecular methods demonstrated the presence of human fecal contamination in 62.5% (4 of 7) of the samples in the coastal waters of Punta Carola Outfall and Playa de los Marinos Berm, San Cristobal (two sites suspected to be impaired during an initial visual assessment). Other sites, such as Laguna de las Ninfas, showed more ephemeral signs of human fecal contamination (Fig. 10.2). This result was not surprising given the location of a nearby submarine sewage discharge pipe at Punta Carola and previous research which has determined fecal contamination in lagoons within or close to population centers in the Galapagos Islands (Lopez and Rueda 2010; Moir and Armijos 2007). The range of *Enterococcus* spp. concentrations is presented as circles in Fig. 10.1, while average concentrations for fecal *Bacteroides* spp., BacHum, and human-specific markers are presented as bar graphs for each site. Concentrations for all assays and all sites averaged $1.38 \times 10^2$, $4.74 \times 10^5$, $1.97 \times 10^3$, and $1.54 \times 10^5$ cell equivalents (CE)/100 ml for *Enterococcus* spp., fecal *Bacteroides* spp., BacHum, and human-specific markers, respectively.

The results of this small-scale study demonstrate concentrations of *Enterococcus* spp. that, according to current water quality standards used globally, could present a risk to public health for those using certain contaminated waters for recreation (World Health Organization 2003). Furthermore, the source-specific molecular marker prevalence and concentrations indicate a strong likelihood that human fecal contamination was present in the nearshore waters surrounding Puerto Baquerizo Moreno and Puerto Ayora during the study period. Based on previous research examining molecular analysis-based levels of *Enterococcus* spp. and Bacteroidales in marine bathing waters (Wade et al. 2010), levels observed at these sites could cause gastroenteritis and respiratory illness rates to exceed 10% for those swimming in these waters. Interestingly, reference clean water beaches located outside urban centers (e.g., Playa Tortuga and Las Tijeretas) showed concentrations that were generally lower for *Enterococcus* spp. and exhibited no apparent signs of human fecal contamination based on the trio of *Bacteroides* spp. markers. Unfortunately, more tourists and locals swim and recreate in areas near towns (e.g., Playa de los Marinos, Playa Carola, and Laguna de las Ninfas), making the potential incidence of disease from waterborne contact higher in these areas.

In 2011 (after this study was conducted), the Puerto Baquerizo Moreno wastewater treatment plant was upgraded for both efficiency and improvements to environmental standards for operation. Prior to this upgrade, the sewage was discharged

**Fig. 10.2** Despite possible human-associated fecal contamination, Laguna de las Ninfas is a popular swimming area for residents of Puerto Ayora, Santa Cruz

into the coastal waters, untreated. The new treatment plant utilizes a chemical treatment, an activated sludge plant with aerobic processing, an aeration system, a disinfection process, and controlled discharge volumes. The plant has the initial capacity to serve 6,000 people, with a second phase of upgrades intended to permit service to be extended to 9,000 people. Our study has not sampled after the sewage upgrade, but future testing should be conducted to verify that there has been an improvement in water quality of the coastal waters proximal to the sewage discharge.

# Recommendations

## *Monitoring*

Consistent monitoring of fecal indicator bacteria and source-specific molecular markers is greatly needed in marine and brackish waters near residential and tourist epicenters of the Galapagos Islands. These monitoring plans should be implemented to include areas of suspected fecal contamination as well as reference (nonimpacted nearshore water comparison) sites. Sampling should include monitoring of *Enterococcus* spp. and more human-specific indicators such as the *Bacteroidales* spp. group. Additional monitoring data such as nutrient analysis for nitrogen and phosphorus, and water quality parameters, such as chlorophyll a, dissolved oxygen, turbidity, total dissolved solids, conductivity, and water temperature, should also be collected.

Initial multi-year monitoring to determine "hot spots" of contamination should be followed by targeted monitoring of contaminated sites. A combined multi-year and "hot spot"-specific approach would permit small-scale variations in contamination, such as weather patterns, and tourist and resident seasonal fluctuations to be detected. Additional changes to the monitoring plan would revolve around alterations (improvements) in wastewater infrastructure, changes in municipal boundaries (sprawl), and epidemiology (e.g., high illness rates associated with a certain marine recreational area). For instance, from January to June, heavy rain showers occasionally occur on the islands, and amounts can quadruple in El Niño years (Adelinet et al. 2008). Though the soils of the Galapagos Islands provide a low runoff potential (Adelinet et al. 2008; d'Ozouville 2008), the presence of intermittent stream beds within Puerto Baquerizo Moreno and Puerto Ayora indicate that runoff can be quite substantial. These potential storms could lead to large overland fecal contaminant flushing events and could be targeted for sampling.

## Improved Infrastructure

Improved water infrastructure is essential to reducing contamination of nearshore marine and brackish waters. Septic systems and unregulated discharges are likely not permitting proper attenuation of fecal microbial pathogens before reaching nearshore waters, due to the subsurface characteristics of the Galapagos Islands. As a result, lagoons such as La Salina, Isabela (Hennessy and McCleary 2011), and Laguna de las Ninfas, Santa Cruz (Lopez and Rueda 2010), or coastal waters such as Playa de los Marinos, San Cristobal, become contaminated. Public wastewater systems, which treat waste before release, similar to the one recently completed at Puerto Baquerizo Moreno, San Cristobal, are needed for the Galapagos Islands population centers.

## Education and Benefits

Though the cost of long-term monitoring and improvements to infrastructure are substantial, the overall benefits are invaluable. Many of the people of the Galapagos acknowledge the importance of water resources and the need for their monitoring, protection, and improvement (Hennessy and McCleary 2011). Continued education and awareness is the key to an engaged citizenry; one which could encourage municipal leaders to invest in improved sanitary infrastructure and water monitoring. This is especially true in developing countries, where the need for clean water and proper sanitation is ultimately up to the residents of the small municipalities (Hagedorn et al. 2011). A portion of the conservation funds that are aimed at protection of the islands and/or tourism income could be allocated for municipal sanitation improvements. Educational institutions (both Ecuadorian and international) could be utilized to assist and implement monitoring plans to reduce costs. International aid groups could also be leveraged, similar to previous studies that were conducted in collaboration between the Galapagos National Park Service and JICA on Santa Cruz Island.

## Conclusions

Investment in water monitoring and infrastructure would benefit the Galapagos Islands on three fronts: through improved quality of life for residents, improved economic benefits (sustained and safe tourism), and improved ecological protection of the rare species for which the Galapagos are famous. Robust microbial contaminant monitoring is needed in nearshore and brackish waters used for recreation (beaches and lagoons) by locals and tourists to better understand the fate and transport of fecal contamination. Additional studies during the "rainy" season should also be conducted to determine inputs of runoff to these waters, and potential seasonally related bacterial reservoir populations and rainfall-associated contamination events. Wastewater infrastructure improvements (centralized wastewater treatment) could dramatically improve water quality in population centers throughout the islands. However, sewage infrastructure improvements must occur in concert with control of human population growth and sprawl to reduce unregulated waste disposal. Education and awareness of the need for clean water from the residents will ultimately influence local government, and infrastructure improvements could be funded utilizing diverse financial resources. Establishing the groundwork for better understanding of the water quality of the Galapagos Islands will inform potential long-range management strategies for improved water quality and marine water resource protection.

**Acknowledgments** The research was made possible through a University of North Carolina Center for Galapagos Studies SEED grant. Special thanks to Veronica Barragan and David Hervas for assistance with sample shipping and Philip Page and Carlos Mena for assistance with logistics. Also, thanks to Monica Green for sample verification and Rodney Gaujardo and Dana Gulbransen for map development. Additional thanks to Parque Nacional Galapagos for assistance with site selection and permission to conduct this research and, particularly, Washington Tapia and Javier López from the Galapagos National Park for assistance at the beginning of the project.

## References

Acevedo-Whitehouse K, Gulland F, Greig D, Amos W (2003) Inbreeding: disease susceptibility in California sea lions. Nature 422:35
Adelinet M, Fortin J, d'Ozouville N (2008) The relationship between hydrodynamic properties and weathering of soils derived from volcanic rocks, Galapagos Islands (Ecuador). Environ Geol 56:45–58

Ahmed W, Goonetilleke A, Powell D, Gardner T (2009) Evaluation of multiple sewage-associated *Bacteroides* PCR markers for sewage pollution tracking. Water Res 43:4872–4877

Al-Bahry SN, Mahmoud IY, Al-Belushi KIA, Elshafie AE, Al-Harthy A, Bakheit CK (2009) Coastal sewage discharge and its impact on fish with reference to antibiotic resistant enteric bacteria and enteric pathogens as bio-indicators of pollution. Chemosphere 77:1534–1539

Armijos E, Palmer M, Moir FC (2002) Selection of an ocean outfall site for the town of Puerto Ayora in the Galapagos Islands. XXVII Congreso Interamericano de Ingenieria Sanitaria y Ambiental, Cancun, Mexico

Atkinson S, Atkinson MJ, Tarrant AM (2003) Estrogens from sewage in coastal marine environments. Environ Health Perspect 111:531–535

Brausch JM, Rand GM (2011) A review of personal care products in the aquatic environment: environmental concentrations and toxicity. Chemosphere 82:1518–1532

Cannicci S, Bartolini F, Dahdouh-Guebas F, Fratini S, Litulo S, Macia A, Mrabu EJ, Penha-Lopes G, Paula J (2009) Effects of urban wastewater on crab and mollusc assemblages in equatorial and subtropical mangroves of East Africa. Estuarine Coastal Shelf Sci 84:305–317

Colford JM Jr, Wade TJ, Schiff KC, Wright CC, Griffith JF, Sandhu SK, Burns S, Sobsey M, Lovelace G, Weisberg SB (2007) Water quality indicators and the risk of illness at beaches with nonpoint sources of fecal contamination. Epidemiology 18:27–35

Converse RR, Blackwood AD, Kirs MA, Griffith JF, Noble RT (2009) Rapid QPCR-based assay for fecal *Bacteroides* spp. as a tool for assessing fecal contamination in recreational waters. Water Res 43:4828–4837

d'Ozouville N (2008) Water resource management in Galapagos: the case of Pelican Bay Watershed. Galapagos report 2007–2008. PNG, FCD and INGALA. Puerto Ayora, Galapagos, pp 158–164

Epler B (2007) Tourism, the economy, population growth, and conservation in Galapagos. Charles Darwin Foundation, Puerto Ayora, Galapagos

Fernandez AR (2008) Coastal nutrient and water budget assessments for Puerto Ayora, Academy Bay. The Pennsylvania State University, Santa Cruz Island. SOARS Report

Fleisher JM, Fleming LE, Solo-Gabriele HM et al (2010) The BEACHES study: health effects and exposures from non-point source microbial contaminants in subtropical recreational marine waters. Int J Epidemiol 39:1291–1298

GNPS, Galapagos National Park Service (2011) Informe de Ingreso de Turistas 2010. Galapagos National Park, Puerto Ayora, Santa Cruz

Hagedorn C, Lepo JE, Hellein KN, Ajidahun AO, Xinqiang L, Li H (2011) Microbial source tracking in China and developing nations. In: Hagedorn C et al (eds) Microbial source tracking: methods, applications, and case studies, Springer, New York

Hennessy E, McCleary AL (2011) Nature's Eden? The production and effects of 'pristine' nature in the Galapagos Islands. Isl Stud J 6:131–156

Islam MS, Tanaka M (2004) Impacts of pollution on coastal and marine ecosystems including coastal and marine fisheries and approach for management: a review and synthesis. Mar Pollut Bull 48:624–649

Jiménez-Uzcátegui G, Milstead B, Márquez C, Zabala J, Buitrón P, Llerena A, Salzar S, Fessl B (2007) Galapagos vertebrates: endangered status and conservation actions. Galapagos report 2006–2007. Charles Darwin Foundation, Puerto Ayora, Galapagos, Ecuador

Kerr S, Cardenas S, Hendy J (2004) Migration and the environment in the Galapagos. Motu Economic and Public Policy Research, Wellington, New Zealand

Kildare BJ, Leutenegger CM, McSwain BS, Bambic DG, Rajal VB, Wuertz S (2007) 16S rRNA-based assays for quantitative detection of universal, human-, cow-, and dog-specific fecal *Bacteroidales*: a Bayesian approach. Water Res 41:3701–3715

Lopez J, Rueda D (2010) Water quality monitoring system in Santa Cruz, San Cristobal, and Isabela. Galapagos report 2009–2010. Charles Darwin Foundation, Puerto Ayora, Galapagos

Moir FC, Armijos E (2007) Integrated water supply and wastewater solutions for the town of Puerto Ayora on the island of Santa Cruz–the Galapagos Islands–Ecuador. Proc Water Environ Feder 70:4816–4842

Noble RT, Blackwood AD, Griffith JF, McGee CD, Weisberg SB (2010) Comparison of rapid quantitative PCR-based and conventional culture-based methods for enumeration of *Enterococcus* spp. and *Escherichia coli* in recreational waters. Appl Environ Microbiol 76:7437–7443

Pearl HW (2009) Controlling eutrophication along the freshwater–marine continuum: dal nutrient (N and P) reductions are essential. Estuar Coast 32:593–601

Penha-Lopes G, Torres P, Narciso L, Cannicci S, Paula J (2009) Comparison of fecundity, embryo loss and fatty acid composition of mangrove crab species in sewage contaminated and pristine mangrove habitats in Mozambique. J Exp Mar Biol Ecol 381:25–32

Rose JB (2006) Identification and characterization of biological risks for water. In: Dura G et al (eds) Management of intentional and accidental water pollution. Springer

Seurinck S, Defoirdt T, Verstraete W, Siciliano SD (2005) Detection and quantification of the human-specific HF183 *Bacteroides* 16S rRNAgenetic marker with real-time PCR for assessment of human faecal pollution in freshwater. Environ Microbiol 7:249–259

Sturm N, Abalos P, Fernandez A, Rodriguez G, Oviedo P, Arroyo V, Retamal P (2011) *Salmonella enterica* in pinnipeds. Chile. Emerg Infect Dis 17:2377–2378

TULAS (2003) Texto Unificado de la Legislación Ambiental Secundaria

Wade TM, Sams E, Brenner KP, Haugland R, Chern E, Beach M, Wymer L, Rankin CC, Love D, Li Q, Noble RT, Dufour AP (2010) Rapidly measured indicators of recreational water quality and swimming-associated illness at marine beaches: a prospective cohort study. Environ Health 9:66

Walsh SJ, McCleary A (2009) Water and health in the Galapagos Islands: a spatial analysis of human-environment interactions. Presentation to UNC institute for global health and infectious diseases, University of North Carolina at Chapel Hill

Walsh SJ, McCleary AL, Heumann BW, Brewington L, Raczkowski EJ, Mena CF (2010) Community expansion and infrastructure development: implications for human health and environmental quality in the Galapagos Islands of Ecuador. J Lat Am Geogr 9:137–159

Watkins G, Cruz F (2007) Galapagos at risk: a socioeconomic analysis of the situation in the Archipelago. Charles Darwin Foundation, Puerto Ayora, Galapagos

Werdeman JL (2006) Effects of populated towns on water quality in neighboring Galapagos bays. Dissertation, University of Washington School of Oceanography, Washington

World Health Organization (2003) Guidelines for safe recreational water environments, vol 1. Feacal pollution and water quality. World Health Organization, Geneva

# Chapter 11
# Research in Agricultural and Urban Areas in Galapagos: A Biological Perspective

Stella de la Torre

## Introduction

Since the creation of the Galapagos National Park in 1959, biological research has greatly contributed to the conservation of the islands and to scientific knowledge in fields like evolutionary biology, taxonomy, biogeography, and population ecology of endemic, native, and introduced species (Parque Nacional Galapagos 2009). However, the ecology of terrestrial ecosystems has been less studied, in particular in agricultural and urban areas.

Conversion of natural ecosystems to agricultural or urban land is the result of a combination of social, economic, and environmental factors that create complex mosaics with different productivity levels, biogeochemical features, and interactions among organisms (Asner et al. 2004). Agricultural and urban areas in Galapagos represent only about 3% of the terrestrial environment but their relevance for the conservation of the islands is unquestionable, since they are the epicenter of human activities that affect natural ecosystems (Caujapé-Castells et al. 2010). The invasion of exotic species, like guava and the goats, began in the agricultural areas of the large islands (ECOLAP and MAE 2007; Itow 2003). The different human activities carried out in these areas have created a matrix of environmental changes that need to be understood to improve the management of protected areas in Galapagos and elsewhere. Studies addressing the effects of human activities on ecosystems are now a priority in research and conservation agendas worldwide (Martino 2001; Prins and Wind 1993), but in the Galapagos these areas of research are still in their beginnings.

S. de la Torre (✉)
College of Biological and Environmental Sciences, Universidad San Francisco de Quito, Quito, Ecuador
e-mail: sdelatorre@usfq.edu.ec

S.J. Walsh and C.F. Mena (eds.), *Science and Conservation in the Galapagos Islands:* *Frameworks & Perspectives*, Social and Ecological Interactions in the Galapagos Islands 1, DOI 10.1007/978-1-4614-5794-7_11, © Springer Science+Business Media, LLC 2013

Terrestrial ecosystems in Galapagos may also be affected by climate change. Studies by the Intergovernmental Panel on Climate Change (IPCC) project a rise of 1.5–4.5°C in the world's mean temperatures in this century and an increase of climatic anomalies such as the El Niño–Southern Oscillation (ENSO) (Houghton et al. 1996; McCarthy et al. 2001). ENSO events in Galapagos are associated with significant rainfall increases and changes in the vegetation cover in terrestrial ecosystems (Robinson and del Pino 1985; Trueman and d'Ozouville 2010). The effects of these future temperature and rainfall increases, as well as of different management strategies, on biological processes may include changes in nutrient dynamics, primary productivity, and the structure of biological communities (Aronson and McNulty 2009; Asner et al. 2004; Bauer et al. 2006; Hollister et al. 2006; Pellens and Garay 1999; Trueman and d'Ozouville 2010). To predict the direction and magnitude of such changes, baseline data should be collected on how nutrients in soil and plants and animal communities vary with land use and ecosystem type, as well as seasonal dynamics in soil nutrients and diversity.

In 2011, long-term research was begun to understand the effects of land use and climate change on the structure and functions of agricultural and urban ecosystems on San Cristobal Island, the second most populated island in the archipelago, with 7,500 inhabitants (ECOLAP and MAE 2007; INEC 2010). Specifically, the aim was to evaluate the effects of land use and climate change on nutrient dynamics, plant productivity, and diversity of animal communities, focusing on soil macroinvertebrates and terrestrial birds. In this chapter, some preliminary results are presented, specifically examining how variability in soil $C/N$ ratio, percent vegetative cover, and diversity of bird and soil macroinvertebrate communities relate to land use.

## Study Areas

In August 2011, four study sites with different land use patterns were selected: urban, organic agriculture, pasture and guava, and restoration sites (Fig. 11.1). The urban site was located near the facilities of GAIAS and the Galapagos Science Center of the Universidad San Francisco de Quito, Ecuador, and the University of North Carolina at Chapel Hill, USA. This site has native, xerophytic vegetation with small trees (e.g., *Bursera graveolens*), shrubs (e.g., *Gossypium darwinii*), and cactus (e.g., *Jasminocereus thouarsii*), as well as some introduced plant species (e.g., *Ricinus communis*). The organic agriculture site was located at Hacienda El Cafetal, near the town El Progreso. Although vegetation in this site is dominated by shrubs of coffee *Coffea* cf. *arabica*, other introduced tree species were also present (e.g., *Cedrela odorata*). Ferns (cf. *Polypodium* sp.) occurred in the undergrowth. The pasture and guava site was located at Hacienda La Tranquila, in the village La Soledad. Vegetation was dominated by introduced plant species, including grasses (e.g., *Paspalum dilatatum*) and trees of guava *Psidium guajava*. The restoration site was also located in Hacienda La Tranquila and was formerly an area of pasture, infested with guava and mora (*Rubus niveus*); few individuals of these two species were still present in the area. The reforested native species included *Lecocarpus darwinii* and

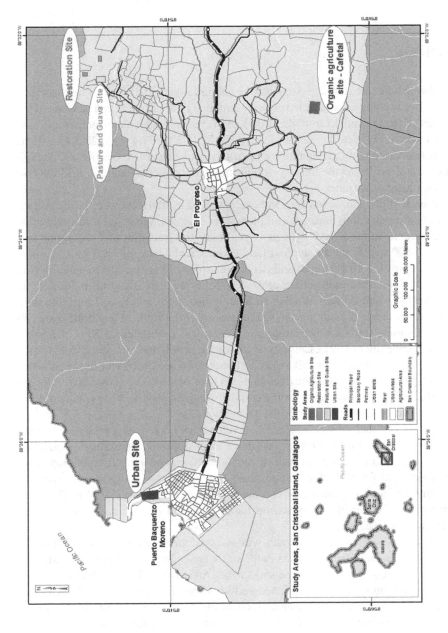

**Fig. 11.1** Location of the study sites, San Cristobal, Galapagos

*Scalesia pedunculata*. Mean linear distance between sites was 4.6 km ± 3.3. The most distant study sites were the urban and restoration sites (linear distance 8 km). The closest sites were the restoration and the pasture and guava sites (0.39 km).

## Methods

Fieldwork was carried out in August 2011 and in January 2012 by 2–3 fieldworkers. These two months were selected as representative of the dry and wet seasons of the islands (Trueman and d'Ozouville 2010). However, although mean temperature and relative humidity were higher in January 2012 (23.2°C–77.4% vs. 26.3°C–79.7%, mean temperature–relative humidity in August and January, respectively), precipitation was zero during the January sample and in the previous month (SEST 840080 Meteorological Station).

In each study site, seven randomly selected 50 m transects were built. In each transect, two randomly located 1 m² plots were selected, separated from each other by at least 10 m for a total of 14 plots per study site (range of plot separation in a site: 10–500 m). From the approximate geometrical center of each plot, one soil sample from 0 to 10 cm depth and two subsamples of the adjacent vegetation (life leaves of all the species inside the plot) were collected once in each climatic season. Soil and leaf samples were dried at ambient temperature, sieved at 2 mm (for soil), and transported to a laboratory in Quito to assay for carbon (C) and nitrogen (N) concentrations. Carbon concentration in leaves was calculated as 50% of organic weight (Schlesinger 1991). Carbon concentration in soil and nitrogen concentration in soil and leaves were directly measured with Walkley-Black and Kjeldahl methods, respectively. For the statistical analyses (see below), the carbon and nitrogen concentrations were averaged for the two leaf samples and the mean concentrations per plot per season were used for the calculations.

In each plot and season, vegetation cover was estimated, as a proxy of primary productivity, using a spherical densiometer. Four different measures of vegetation cover were performed, one in each cardinal direction, and a mean vegetation cover was calculated for each climatic season. A rate of change of vegetation cover was computed by dividing the mean percentage of cover in the wet season by the mean percentage of cover in the dry season for each plot to include this variable in the statistical analyses (see below).

The diversity of soil invertebrates was assessed through surveys of two subplots of 25 cm² in each of the 1 m² plots in the study areas. In each subplot, 2–4 different surveys were conducted, from the soil surface to 5 cm depth, in each climatic season. Invertebrates were photographed and identified for their taxonomic order; no specimens were collected. Shannon diversity indices (Smith and Smith 2000) were calculated with the number of orders and the number of individuals in each order, found in each survey for each subplot. For the statistical analyses (see below), the indices of the two subplots were averaged and the mean index per plot per season was used for the calculations.

To assess bird diversity, three fixed observation points were selected along the transect system in each study site. We carried out 2–4 30 min censuses, from 0600 to 0800 h and from 1600 to 1800 h, in each observation point in the dry and rainy seasons. In the censuses, bird species actively using the area around the observation point, within a 30 m radius, were recorded. Birds were identified with field guides. Occasionally (less than 20% of all surveys), we could not identify ground finches to the species level and recorded them as *Geospiza* sp. In even fewer cases (less than 5% of all surveys), the species could not be identified and we recorded those individuals as "not indentified." Shannon diversity indices were calculated for each survey, including the *Geospiza* sp. and the "not identified" bird categories in those surveys with identification problems.

## Quantitative Analyses

Repeated-measures multifactorial ANOVAs were carried out to compare the ecosystem variables among sites in both climatic seasons: transformed (arcsin sqrt (p)) percentages of carbon and nitrogen in soil and leaf samples, $C/N$ ratios in soil, transformed percentages of vegetation cover, and diversity indices of soil macroinvertebrates. This model was selected since measurements for all these variables were taken from the same plots in each season. A multifactorial ANOVA was carried out to compare diversity indices of birds among sites and between seasons. A one-way ANOVA was used to compare the transformed (sqrt (p)) wet/dry rate of the change of vegetation cover among sites.

Simple linear regressions were carried out to evaluate the influence of carbon and nitrogen concentration, as well as of the soil $C/N$ ratio on vegetation cover and diversity of soil macroinvertebrates in both climatic seasons; transformed variables were used for the calculations when appropriate. Increased available nitrogen in soil may increase primary productivity and vegetation cover (Galloway et al. 2003), whereas increased $C/N$ ratios in soil may reduce decomposition rates (Ordoñez 2010), thereby impacting the community dynamics of soil invertebrates. Considering that some invertebrates may be prey for most bird species (see Abott et al. 1977), a Pearson correlation was calculated between the Shannon diversity indices of soil macroinvertebrates and birds across seasons.

## Results

The study sites differ significantly in nitrogen and soil concentration in soil and leaves (see below). The restoration site showed the highest nitrogen and carbon concentrations in soil, whereas the pasture and guava site had the lowest concentrations of these two elements in soil in both climatic seasons ($N/F_{3,52} = 4.45$, $p = 0.07$; $C/F_{3,52} = 3.13$, $p = 0.033$) (Figs. 11.2 and 11.3, Table 11.1).

The highest nitrogen concentration in leaves was found in the organic agriculture site in both seasons, whereas the lowest was recorded in the pasture and guava site

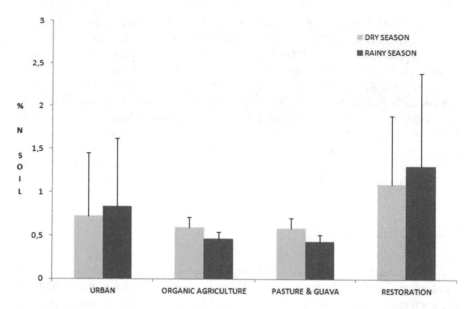

**Fig. 11.2** Mean percentage (± standard deviation) of the percentage of nitrogen in soil in the four study sites in the dry and rainy season samples

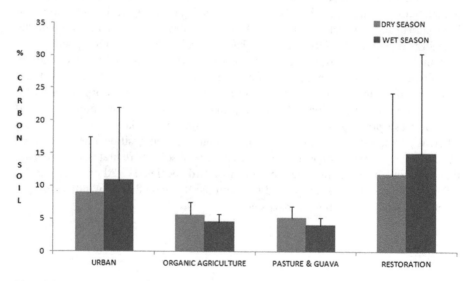

**Fig. 11.3** Mean percentage (± standard deviation) of the percentage of carbon in soil in the four study sites in the dry and rainy season samples

($F_{3,52}=44.76, p<0.0001$). Nitrogen concentrations in leaves were significantly higher in the wet season in all sites ($F_{1,52}=45.76, p<0.0001$) (Fig. 11.4, Table 11.1).

On the other hand, the pasture and guava site had the highest carbon concentration in leaves, whereas the lowest concentration was found in the urban site

**Table 11.1** Mean values (± standard deviation) of the studied variables in the four study sites (D, dry season; R, rainy season)

| Variable | Urban | | Organic agriculture | | Pasture and guava | | Restoration | |
|---|---|---|---|---|---|---|---|---|
| | D | R | D | R | D | R | D | R |
| % Nitrogen soil | 0.72±0.73 | 0.83±0.79 | 0.59±0.12 | 0.46±0.08 | 0.58±0.12 | 0.43±0.08 | 1.09±0.80 | 1.3±1.09 |
| % Carbon soil | 8.95±8.53 | 10.85±11.14 | 5.56±1.93 | 4.54±1.15 | 5.16±1.73 | 4.03±1.14 | 11.78±12.50 | 15.00±15.29 |
| % Nitrogen leaves | 1.87±0.43 | 2.55±0.72 | 2.82±0.25 | 3.60±0.47 | 1.63±0.30 | 2.28±0.25 | 2.30±0.47 | 2.78±0.56 |
| % Carbon leaves | 41.68±2.10 | 37.74±2.22 | 44.80±0.61 | 41.68±2.10 | 45.86±0.43 | 43.10±1.38 | 45.34±0.30 | 38.72±1.02 |
| C/N soil | 11.33±5.53 | 12.04±5.83 | 9.24±1.64 | 9.85±1.75 | 8.78±1.17 | 9.34±1.28 | 9.68±3.64 | 10.28±3.82 |
| % Vegetation cover | 12.02±5.51 | 12.28±6.40 | 87.42±16.94 | 64.45±2.70 | 25.7±32.50 | 22.7±9.30 | 22.42±14.70 | 14.52±10.38 |
| H macroinvertebrates | 0.69±0.26 | 0.78±0.39 | 1.03±0.27 | 1.53±0.22 | 1.33±0.44 | 0.92±0.27 | 1.10±0.32 | 1.14±0.24 |
| H birds | 1.17±0.26 | 1.37±0.26 | 1.11±0.26 | 1.44±0.31 | 1.45±0.20 | 1.27±0.23 | 1.47±0.34 | 1.54±0.44 |

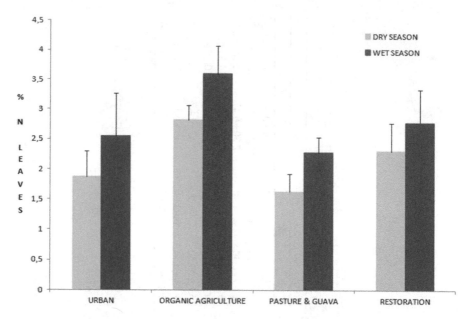

**Fig. 11.4** Mean percentage (± standard deviation) of the percentage of nitrogen in leaves in the four study sites in the dry and rainy season samples

($F_{3,52}=29.99$, $p<0.0001$). Carbon concentrations in leaves were significantly lower in the wet season in all sites ($F_{1,52}=425.3$, $p<0.0001$) and there was a significant interaction between site and season ($F_{3,52}=24.94$, $p<0.0001$), suggesting the strong influence of climate on this variable (Fig. 11.5, Table 11.1).

No significant differences were found in the $C/N$ ratios in soil among sites, but $C/N$ ratios in soil were significantly higher in the wet season in all sites ($F_{1,52}=293.4$, $p<0.0001$) (Table 11.1).

Vegetation cover was significantly denser in the organic agriculture site, whereas the most sparse cover occurred in the urban site in both seasons ($F_{3,52}=33.90$, $p<0.0001$). Vegetation cover was significantly denser in the dry season study period in all sites ($F_{1,52}=7.62$, $p=0.0079$) and there was a significant interaction between site and season ($F_{3,52}=3.79$, $p=0.0018$) (Fig. 11.6, Table 11.1). No significant differences were found in the rate of change of vegetation cover among sites.

A total of 14 orders of soil invertebrates were recorded in the dry and rainy season samples in the four study sites. Gastropoda (snails), Diplopoda (millipedes), Isopoda (pill bugs), and Haplotaxida (earthworms) were frequently recorded. Significant differences were found in the diversity indices among sites; the lowest diversity occurred in the urban site in both seasons. The highest diversity in the dry season was found in the pasture and guava site, whereas in the rainy season the highest diversity was found in the organic agriculture site ($F_{3,52}=14.25$, $p<0.0001$). There was also a significant interaction between site and season since in the rainy season increase in diversity did not occur in all sites ($F_{3,52}=11.3$, $p<0.0001$) (Fig. 11.7, Table 11.1).

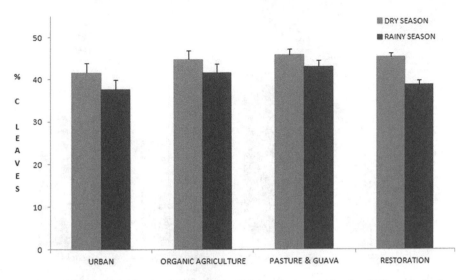

**Fig. 11.5**  Mean percentage (± standard deviation) of the percentage of carbon in leaves in the four study sites in the dry and rainy season samples

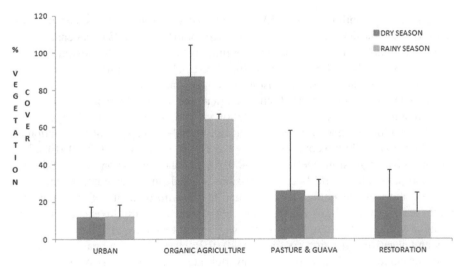

**Fig. 11.6**  Mean percentage (± standard deviation) of the percentage of vegetation cover in the four study sites in the dry and rainy season samples

A total of 11 bird species were recorded in the censuses in both seasons; two of them were exotic species, the smooth-billed ani *Crotophaga ani* and the cattle egret *Bubulcus ibis* (Appendix 1). The smooth-billed ani was recorded in all the study sites, but the majority of recordings were obtained in the organic agriculture site (1.58 observations per census) and in the pasture and guava site (1.23 obs./census), both in the dry season. Observations of the cattle egret were only made in the pasture

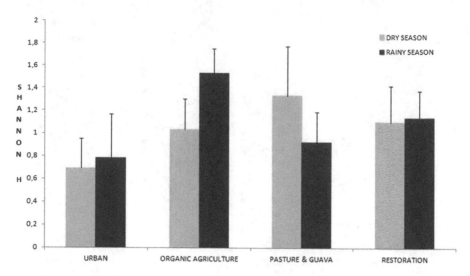

**Fig. 11.7** Mean percentage (± standard deviation) of the Shannon biodiversity indices (H) of soil macroinvertebrates in the four study sites in the dry and rainy season samples

and guava site in both seasons (1.63 obs./census dry season, 0.17 obs./census rainy season) and in the dry season censuses of the restoration site (0.25 obs./census). The highest diversity of the terrestrial bird community was found in the restoration site in both seasons (H=1.46±0.34 dry season, 1.54±0.44 rainy season), but the differences among sites and between seasons were not significant (Table 11.1).

The linear regressions carried out to evaluate the relation between nutrient concentration in soils and leaves with vegetation cover and rate of cover change across all sites had low $R^2$s and were not significant, with the exception of the relation of the $C/N$ ratio in soil with cover in the wet season that showed a low $R^2$ (0.18), a low and negative regression coefficient (−0.015), but the regression was significant ($p=0.001$). Similarly, the linear regressions carried out to evaluate the relation between nutrient concentration in soils and leaves with diversity of macroinvertebrates across sites had low $R^2$s and were not significant, with the exception of the relation with the $C/N$ ratio in soil in the wet season that showed a low $R^2$ (0.12), a low and negative regression coefficient (−0.019), but it was significant ($p=0.008$). The correlation coefficient between invertebrate and bird diversity across sites ($r=0.57$) was not significant.

## Discussion

Although preliminary, the significant differences found in some of the studied ecological variables among sites suggest that the land use patterns have considerable effects on the structure and function of terrestrial ecosystems in the Galapagos.

However, similar analyses have not yet been conducted in "control" sites with native ecosystems to correctly assess the magnitude of the changes. Such analyses will begin in the second year of research and may help to better explain the observed differences.

The results provide evidence that pasture, in combination with guava, affects nutrient availability. The difference between the highest concentrations of nitrogen and carbon in the restoration site and the concentrations of both nutrients in the pasture and guava site was almost twofold in both seasons. The difference is even more remarkable considering that both sites are separated by only about 0.4 km and that five years ago the restoration site was also a pasture area (G. Sarigu, personal communication).

Nitrogen scarcity may be related to the very low concentration of this nutrient in the leaf samples from the pasture and guava site, as has been reported in other studies (van Arendonk et al. 1997; Ordoñez 2010). However, all the analyses do not permit a determination of the limiting factors affecting vegetation cover as an indicator of primary productivity, and the diversity of animal communities, specifically of soil macroinvertebrates and terrestrial birds.

Results suggest that, at least in the rainy season, vegetation cover and macroinvertebrate diversity are partially and negatively related to the $C/N$ ratio in soil. This ratio is considered to be an indicator of the quality of leaf litter for decomposers, affecting the decomposition rates and the nitrogen supply for plants. High $C/N$ ratios are related to less availability of nitrogen for plants since most of the nitrogen is assimilated by the decomposers (Alvarez-Sánchez 2001). This may explain the negative relationship between this ratio and vegetation cover found in the study areas. Higher macroinvertebrate diversity in areas with lower $C/N$ ratios could be expected, but coverage and diversity patterns may also be influenced by other factors not related to nutrient supplies.

Human intervention, through selective cutting and pruning, for example, affected vegetation cover in all the study sites. The lower percentages obtained in the rainy season samples in some of the plots were caused by the previous cutting of vegetation in all sites by landowners for different reasons that could not be controlled in the study. The lowest values of the vegetation cover in the urban site are certainly also related to the drier conditions in the coastal zones of the islands (Trueman and d'Ozouville 2010).

Human influence on patterns of macroinvertebrate diversity may occur through the eventual use of pesticides, although we did not witness their use in the field, or indirectly by affecting vegetation cover and soil characteristics. In the pasture and guava site, for example, just before the rainy season, samples were occupied by cows and horses that ate a large portion of the plants and compacted the soil. Data collection across several years may help to better evaluate the influence of these other variables that could not be controlled in this study, provided we are able to record their occurrence adequately. This could be achieved by increasing the participation of local people in this research. These chrono-sequences may also provide us with insight into ecosystems' resilience and the impact of current climate change.

Although not significant, differences in bird diversity among sites point to the importance of native vegetation in the diversity of bird species. The site with the highest diversity indices in both seasons was the restoration site and, although diversity included records of two introduced bird species, the number of observations of these species was lower than in the other sites.

To my knowledge, this is the first study to analyze the effects of land use and climate on nutrient dynamics and community diversity in agricultural systems in the Galapagos. Evidently, more data and analyses are needed to understand the direction and extent of the impact of land use and climate changes on island ecosystems. Some of the ideas for future work (e.g., including protected areas with native vegetation as control sites) will be carried out in the short- to midterm, whereas others (e.g., chrono-sequences) may require collaboration with other researchers.

**Acknowledgments** This research was funded by a grant from GAIAS—Universidad San Francisco de Quito. A research permit (PC-30-11) was granted by the Galapagos National Park that allowed this research to be conducted. I deeply acknowledge Isabel Villarruel for her outstanding assistance in the fieldwork and data analyses; Pablo Yépez for his support in all the stages of the study; Aaron Moody for his valuable comments on this manuscript; and Carlos Mena and Steve Walsh of the Galapagos Science Center (USFQ/UNC) for their valuable support during the fieldwork and for inviting me to participate in this book. Geovanny Sarigu from Hacienda La Tranquila and Nicolás Balón and Edgar Román from Hacienda El Cafetal kindly allowed me to carry out the field research in those areas and constantly supported our fieldwork. My special thanks to Sofía Tacle, Leandro Vaca, Courtney Butnor, Cecibel Narváez, María Angélica Moreano, Máximo and Marlene Ochoa, and all the personnel from GAIAS and the Galapagos Science Center for their support in the field.

## Appendix A. Appendix 1. List of bird species recorded in each study site

| Scientific name | Urban | Organic agriculture | Pasture and guava | Restoration |
|---|---|---|---|---|
| *Crotophaga ani* | X | X | X | X |
| *Nesomimus melanotis* | X | X | X | X |
| *Dendroica petechia* | X | X | X | X |
| *Certhidea olivacea* | | X | X | X |
| *Geospiza fortis* | X | X | X | X |
| *Geospiza fuliginosa* | X | X | X | X |
| *Platyspiza crassirostris* | | X | X | X |
| *Camarhynchus pallidus* | | X | X | X |
| *Camarhynchus parvulus* | | X | | X |
| *Myiarchus magnirostris* | X | | | |
| *Bubulcus ibis* | | | X | X |

# References

Abott I, Abbot LK, Grant PR (1977) Comparative ecology of Galapagos ground finches (*Geospiza Gould*): evaluation of the importance of floristic diversity and interspecific competition. Ecol Monogr 47:151–184

Alvarez-Sánchez J (2001) Descomposición y ciclos de nutrientes en ecosistemas terrestres de México. Acta Zoológ Mexic 1:11–27

Aronson EL, McNulty SG (2009) Appropriate experimental ecosystem warming methods by ecosystem, objective, and practicality. Agr Forest Meteorol 149:1791–1799

Asner GP, Townsend AR, Bustamante MM, Nardoto GB, Olander LP (2004) Pasture degradation in the central Amazon: linking changes in carbon and nutrient cycling with remote sensing. Glob Chang Biol 10:844–862

Bauer IE, Apps MJ, Bhatti JS, Lal R (2006) Climate change and terrestrial ecosystem management: knowledge gaps and research needs. In: Bhatti JS, Lal R, Apps MJ, Price MA (eds) Climate change and managed ecosystems. CRC Press, Boca Raton

Caujapé-Castells J, Tye A, Crawford DJ, Santos-Guerra A, Sakai A, Beaver K, Lobin W, VincentFlorens FB, Moura M, Jardim R, Gómes I, Kueffer C (2010) Conservation of oceanic island floras: present and future global challenges. Perspect Plant Ecol Evol Systemat 12:107–129

ECOLAP, MAE (2007) Guía del Patrimonio de Áreas Naturales Protegidas del Ecuador. Ecofund, FAN, DarwinNet, IGM, Quito

Galloway JN, Aber JD, Willem Erisman J, Seitzinger SP, Howarth RW, Cowling EB, Cosby J (2003) The nitrogen cascade. Bioscience 53:341–356

Hollister RD, Webbert PJ, Nelson FE, Tweedie CE (2006) Soil thaw and temperature response to air warming varies by plant community: results from an open-top chamber experiment in Northern Alaska. Artic Antarct Alpine Res 38:206–215

Houghton JT, Meira Filho LG, Bruce J, Lee H, Callender BA, Haites E, Harris N, Maskell K (eds) (1995). Climate change 1995: the science of climate change. Cambridge University Press, Cambridge

INEC (2010) Resultados censo de población 2010. http://www.inec.gob.ec (accessed 14 June 2012).

Itow S (2003) Zonation pattern, succession process and invasion by aliens in species-poor insular vegetation of the Galapagos Islands. Glob Environ Res 7:39–57

McCarthy JJ, Canziani OF, Leary NA, Dokken DJ, White KS (eds) (2001) Climate change 2001: impacts, adaptation, and vulnerability. Cambridge University Press, Cambridge

Martino D (2001) Buffer zones around protected areas: a brief literature review. Electron Green J. 1:15. http://www.escholarship.org/uc/item/02n4v17n (accessed 11 Oct 2010).

Ordoñez JC (2010) Environmental filtering vs. natural variation and plant strategies: key components of plant trait modulation by nutrient supply. Thesis 2010–01 of the Institute of Ecological Science, Vrije Universiteit, Amsterdam.

Parque Nacional Galapagos (2009) Galapagos, cincuenta años de ciencia y conservación. De Roy T (ed) Imprenta Mariscal, Quito. http://www.Galapagospark.org/nophprg.php?page=ciencia_investigacion_proyectos.

Pellens R, Garay I (1999) Edaphic macroarthropod communities in fast-growing plantations of *Eucalyptus grandis* Hill ex Maid (Myrtaceae) and *Acacia mangium* Wild (Leguminosae) in Brazil. Eur J Soil Biol 35:77–89

Prins H, Wind J (1993) Research for nature conservation in south-east Asia. Biol Conserv 63:43–46

Robinson G, del Pino EM (1985) El Niño en las islas Galapagos, el evento de 1982–1983. Fundación Charles Darwin, Quito

SEST 840080 meteorological station www.tutiempo.net/clima/San_Cristóbal_Galapagos/08-2011/840080.htm (accessed 23 Apr 2012)

Schlesinger WH (1991) Biogeochemistry, an analysis of global change. Academic Press, New York

Smith RL, Smith TH (2000) Elements of ecology, 4th edn. Addison Wesley Longman, San Francisco

Trueman M, d'Ozouville N (2010) Characterizing the Galapagos terrestrial climate in the face of global climate change. Galapagos Res 67:26–37

Van Arendonk JCM, Niemann GJ, Boon JJ, Lambers H (1997) Effects of nitrogen supply on the anatomy and chemical composition of leaves of four grass species belonging to the genus *Poa*, as determined by image-processing analysis and pyrolysis mass spectrometry. Plant Cell Environ 20:881–897

# Chapter 12
# A Geographical Approach to Optimization of Response to Invasive Species

George P. Malanson and Stephen J. Walsh

## Introduction

Invasive species have been identified as one of the major ecological consequences of globalization or as themselves a major global change on the level of climate change or land use change (e.g., Mack et al. 2000; Hobbs and Mooney 2005; Sharma et al. 2005; Ricciardi 2007). While current rates of invasive species movement may have occurred locally in the past, the global extent of the current phenomenon is unprecedented (Mooney and Hobbs 2000). The potential impacts are also high; Parker et al. (1999) identified impacts on native species at the individual, population (both genetic and dynamics), community, and ecosystem process levels and Levine et al. (2003) reviewed 150 studies to evaluate the processes of impact. As these impacts increase, the potential exists for positive feedbacks that could accelerate invasions (Simberloff and Von Holle 1999). The overall problem is well established, and estimates of the economic impact range into the billions of US dollars (over $100 billion annually in the USA alone according to NRC (2002), over £239 million in Britain [White and Harris 2002]) or are judged "incalculable" (Mack et al. 2000). The potential to drive native species to extinction is one incalculable cost (cf. Mooney and Cleland 2001; Stockwell et al. 2003).

   Much of the concern and impact of invasive species comes from observations on islands (cf. Bergstrom and Chown 1999). In his classic work on island biogeography, Carlqist (1965) identified introduced animals as the "chief plague" affecting native species, and MacArthur and Wilson (1967) identified invasibility as a characteristic

G.P. Malanson (✉)
Department of Geography, University of Iowa, Iowa City, IA 52242, USA
e-mail: george-malanson@uiowa.edu

S.J. Walsh
Department of Geography, Center for Galapagos Studies, University of North Carolina
at Chapel Hill, NC 27599, USA
e-mail: swalsh@email.unc.edu

S.J. Walsh and C.F. Mena (eds.), *Science and Conservation in the Galapagos Islands:*       199
*Frameworks & Perspectives*, Social and Ecological Interactions in the Galapagos Islands 1,
DOI 10.1007/978-1-4614-5794-7_12, © Springer Science+Business Media, LLC 2013

**Fig. 12.1** Eradication of invasive ornamentals in Puerto Villamil

of islands. Extensive studies on invasive species have been carried out in Hawaii (e.g., Loope and Mueller-Dombois 1989), New Zealand (e.g., Wiser et al. 1998), Juan Fernández (Dirnbock et al. 2003), and the super island of Australia (e.g., Hobbs 2001), among others. Concerns are heightened on islands because of the high proportion of endemics (e.g., Porter 1976, 1979; Tye and Francisco-Ortega 2011).

## In the Galapagos

Invasive species are a definite threat to the native species, biogeographical uniqueness, and scientific heritage of the Galapagos Islands (MacFarland and Cifuentes 1996). Hamann (1984), in a chapter on "Changes and Threats to the Vegetation," cited just alien plants and animals. Mauchamp (1997) reported that 550 native plant species were faced with 438 alien species, of which he categorized 11 as invasive/ aggressive. Tye (2006) and Tye and Francisco-Ortega (2011) extended this study to report 552 native species, 58 questionably native, and 486 aliens. The small proportion of invasive/aggressive is not indicative of the potential areal extent or damage (Mauchamp et al. 1998). Itow (2003) noted that invasives may have an advantage because the Galapagos are poor in tree species.

Baseline inventories of alien species are also underway (Charles Darwin Foundation, http://www.galapagos.org/2008/index.php?id=60). The agricultural and urban zones of the inhabited islands have been surveyed, and the results indicate that ongoing monitoring and eradication efforts are needed as new species and/ or populations are found (Fig. 12.1). The Charles Darwin Research Station (CDRS) identified "prioritization" as an important challenge in invasive species research and management. A major focus is on species biology in order to identify those most likely to spread, but attention to identifying places to concentrate efforts to minimize spread or eradicate populations is needed.

Hamann (1984, 1991) and Itow (2003) identified four invasive plant species of concern: *Psidium guajava* (guava), *Cinchona succirubra* (quinine), *Lantana camara*, and *Pennisetum purpureum*. The major invasive species of concern are now guava and *Rubus niveus* (blackberry, locally called mora, along with four other *Rubus* species).

## Guava (According to Hamman (1991), Introduced to the Islands in 1858)

Guava is a shrub or small tree widely cultivated for its edible fruit (Ellshoff et al. 1995). Eckhardt (1972) gave it special attention and reported that it was strongly established in dense thickets on some of the larger islands. Its extensive, problematic character was reiterated by Schofield (1989). In the Galapagos, it is considered a transformer species (Tye 2002), an invasive species that changes "the character, condition, form, or nature of ecosystems over substantial areas relative to the extent of that ecosystem" (Richardson et al. 2000). Guava generally thrives in pasture and other grasslands, roadsides, cropland, and other disturbed areas (GISD 2005), forming dense thickets that prevent regeneration of native vegetation and reduce species richness. In the Galapagos Islands, the species can also invade natural forests (Binggeli et al. 1998). In the Galapagos, both the high rainfall of El Niño and fire are thought to have hastened its spread (Tye 2001). It was introduced into the humid highlands on Isabela Island (Fig. 12.2), and it is now extensive through the agricultural zones from which it has expanded into the surrounding park land that originally consisted of Scalesia forests and the treeless fern-sedge zone (Hamann 1981).

## Blackberry (According to Itow (2003), First Observed in 1983)

Blackberry is a shrub that is well known as a genus over much of the temperate world. *Rubus niveus* is one of the few blackberry species that grows in the tropics, but it is an aggressive invasive in Hawaii and the Galapagos. It grows best in partial shade and can form nearly impenetrable thickets up to 4 m (but usually 2 m) tall (Renteria et al. 2007; St Quinton et al. 2011). While generally eradicated in the Isabela, San Cristobal, and Floreana highlands, it could reemerge if there is a decay of management programs related to land abandonment. On Santa Cruz Island, it remains a significant problem.

**Fig. 12.2** Guava (*Psidium guajava*) covering a slope next to agricultural clearing, Isabela

## *Elephant Grass*

Elephant grass (*Pennisetum purpureum*), most abundant in pastures of the lower highlands and along road edges in disturbed areas, is a tall grass, up to 2 m in height. It can be used for animal fodder, hence its introduction to the islands. It is, however, invasive and can exclude other plant species over wide areas (cf. Space and Falanruw 1999; Space and Flynn 2000 for other Pacific islands). Our observations indicate that it is extensive along roads but it also covers whole fields.

## *Quinine (First Arrival in 1946 (Itow 2003))*

Itow (2003) cited the low number of tree species in the Galapagos as a reason for the invasive success of this tree. It is a threat to *Scalesia* forests on Santa Cruz, where it is currently restricted. It is a target for control measures (Buddenhagen et al. 2004; Jäger and Kowarik 2010a, b).

## Recent Research

The botany program at the CDRS has a strong program of research into both the native and alien plant species on the islands (e.g., http://www.youtube.com/watch?v=kjO8LPsAcvs; thorough review by Tye and Francisco-Ortega 2011). Ongoing work is extending baseline surveys of the flora of the Galapagos to a complete areal survey of all islands. Guezou et al. (2010) reported on a survey of the four inhabited islands and identified 257 new alien species; some of these they eradicated as they went (personal observation). New species, populations of species thought to be extinct, and additional populations are being discovered (Tye and Francisco-Ortega 2011). Populations of endangered species are being monitored to the extent feasible. A variety of studies on management and restoration are underway, some linked to the successful eradication of alien animal species, particularly goats.

Much of the recently published research on invasive species is about animals (e.g., Renteria and Buddenhagen 2006; Atkinson et al. 2008; Brand et al. 2012), where successful but costly programs to eradicate goats are partially completed (e.g., Carrion et al. 2011), but work on plants is increasing (some related to recovery from goats; Hamman 1979, 1993, 2004). Wilkinson et al. (2005) modeled possible forest restoration in the context of invasive trees (primarily *Quinine*); Buddenhagen and Jewell (2006) reported on seed viability; Castillo et al. (2007) reported drought tolerances for native and nonnative lantana; Jäger et al. (2007, 2009) reported on how quinine transforms the ecosystem it invades, while Jäger and Kowarik (2010a, b) reported on the recovery of the ecosystem following manual eradication. Watson et al. (2010) assessed the overall degradation of habitat primarily due to invasive species, and Trueman et al. (2010) studied both the plant and human characteristics of plant invasion, differentiating naturalized, invasive, and transformer species; as a control measure, they specified intensified interisland quarantine, but Mireya Guerrero and Tye (2011) noted that some interisland dispersal is in the guts of native bird species (cf. Heleno et al. 2011). These studies are set in a context of ongoing work on the native flora (e.g., Coffey et al. 2011). More recently, the web of connections with the human population of the Galapagos has been explored (Guezou et al. 2007; Gonzalez et al. 2008; Miller et al. 2010; cf. MacFarland and Cifuentes 1996).

## Spatial Optimization

Optimization of effort is not new in the Galapagos (e.g., Cruz et al. 2009). Spatial optimization in conservation science has focused primarily on the covering problem, i.e., what arrangement of land uses would conserve the most species or other environmental good (e.g., Church and Gerrard 2003; Matisziw and Murray 2006; cf. Kupfer et al. 2006). Here, we are interested in optimizing spatial coverage in a different sense. We discuss the problems and potential approaches to improve the detection and eradication of invasive species, so we are interested in processes more like spatial searches. We examine two complementary approaches: remote sensing and fieldwork (Rew et al. 2005). We will also focus on additional constraints that apply in island situations such as the Galapagos. Lavoie et al. (2007) illustrated how geographic methodologies could be applied in the case of goat eradication; they used GIS, GPS, and remote sensing to plan and study eradication methods, including ground- and helicopter-based hunting, but did not optimize per se.

Detection and eradication are essentially spatial problems. They primarily require learning where the invasives are and getting there. Additional problems of eradication are manifest, however, and involve lack of success, impacts of the eradication process on the environment, and impacts—at least initially—of the loss of the function of the invasive species.

Islands (and potentially other isolates) present special spatial problems for the detection and eradication of invasive species. From the perspective of remote sensing and field work, respectively, a major optimization issue boils down to the amount of time spent over or on the ocean. Additional decisions that can be made to improve the process apply in many situations but may be magnified for islands, at least for the Galapagos, and we will discuss some of these as well.

# Remote Sensing

The problem for remote sensing is that it is desirable to detect invasive plant species at the earliest stage possible, i.e., when they are smaller than the spatial resolution of the most readily available satellite imagery and may also be similar in spectral signature to other vegetation. Walsh et al. (2008) illustrated how this problem could be attacked using a combination of higher-spatial resolution (Quickbird) and higher-spectral resolution (Hyperion) imagery, but they were looking at well-established guava.

Several studies using hyper-spectral, over multispectral, data report considerable success with invasive studies. For example, Underwood et al. (2003), using AVIRIS hyper-spectral imagery, showed that the invasive weeds ice plant and jubata grass could be mapped with an accuracy of 94% and 89%, respectively, using band ratios and continuum removal. Continuum removal is a de-correlation technique that maximizes the effects of spectral absorption features. Mundt et al. (2005), using HyMap imagery, showed similar accuracies when mapping hoary cress with at least 30% cover using the Minimum Noise Fraction (MNF) (Green et al. 1988) and the Mixture-Tuned Matched Filter (MTMF) method (Boardman 1993; Boardman et al. 1995). Williams and Hunt (2002), using AVIRIS imagery, showed that the MTMF method produced the most accurate results for mapping leafy spurge, while Miao et al. (2006), using CASI-2 hyper-spectral imagery, showed that linear unmixing produced high accuracies for yellow starthistle. Hyper-spectral remote sensing has been used to assess the spectra of blackberry and to effectively quantify its distribution in open canopies (Dehaan et al. 2007).

To identify the advancing edge of guava as it colonizes new territories, a remote-sensing system, such as WorldView-2 (WV-2) from Digital Globe, must be capable of distinguishing young plants from the background matrix and characterizing low-density patches before they coalesce into an open or broken canopy. WV-2, launched in 2009, operates in the 450–800 nm spectral range for the panchromatic imaging mode and 400–900 nm for the multispectral imaging mode. Simultaneously collected, the panchromatic mode has a spatial resolution of 0.46 m, while the multispectral mode has a spatial resolution of 1.84 m. Advances of this system over prior high-spatial-resolution systems for vegetation characterizations is the yellow channel (585–625 nm), red edge channel (705–745 nm), and a near-infrared channel (860–1,040 nm). It also has a "coastal" channel for studies of bathymetry based upon its chlorophyll and water-penetrating capacity.

But even with high-spatial-resolution, satellite systems, and the possible fusion of multiple spectral, spatial, and temporal-resolutions data sets, a persistent challenge is the capacity to map sub-canopy vegetation, such as blackberry. Increasingly, high-spatial-resolution LIDAR systems are being used to map multiple plant canopies, using first and last return system as well as continuous return systems through aircraft platforms (Cracknell and Hayes 2007).

To detect less extensive invasions of smaller individual plants, it might be necessary to use multispectral and even hyper-spectral imagery at spatial resolutions that cannot be achieved with satellite platforms. Aircraft platforms become expensive, however, and here is where the island versus ocean coverage must be considered. The first question will be a choice of platforms and bases either on the mainland or the islands, with airports on Baltra, San Cristobal, and Isabela. Then, optimal flight plans will maximize the time spent over the islands and areas of islands that most need observation, with less time spent over other island areas and minimizing time over the ocean. This choice becomes the old traveling salesman problem of economic geography because it is the interisland distance that must be minimized, given the constraint of a starting airfield. The choice of island and area is discussed further below.

But while the use of satellite systems have dominated over the last 15 years as spatial and spectral resolutions have improved (e.g., Everitt et al. 1995; Lass et al. 2005; Hunt and Parker-Williams 2006), the use of color infrared, natural color, and panchromatic aerial photography continues to be an important technology to map the localized occurrences and spread of invasive plants in a host of settings (e.g., Maheu-Giroux and de Blois 2005, Ge et al. 2006). Aerial photography has been used to detect and monitor invasive plants generally using large- to moderate-scale vertical, stereoscopic aerial photography collected initially in analog form and/or converted to digital data for subsequent analysis (Naylor et al. 2005). Image classification has been traditionally used to map land use/land cover with an emphasis on the condition and pattern of invasive plant species and occasionally to assess space-time trends and perspectives of plant invasion using historical imagery (Ge et al. 2006). Among the analytical approaches for feature extraction and enhancement, spatial filters have been generated for windows ranging from $3 \times 3$ to $9 \times 9$ pixels to assess image texture (e.g., homogeneity, contrast, dissimilarity) on aerial photography. Vegetation indices have also been derived to assess plant conditions to further distinguish invasive plants from the background matrix (Lass et al. 2005; Ge et al. 2006). Most recently in the Galapagos, the March 2007 aircraft mission characterized the islands of the archipelago at a scale of 1:30,000, in natural color, and with standard forward- and side-lap for stereoscopic viewing.

## Field Surveys

The need to improve on-the-ground efforts for invasive species has been recognized elsewhere. For example, although titled eradication, Regan et al. (2006) presented a study of how to find the stopping rule for monitoring (stopping when eradication is calculated) by comparing the costs of continuing versus those that would be incurred by a renewed invasion (at a calculated probability). Other optimization models have focused on eradication or biological control. DeWalt (2006) simulated the population dynamics of an invasive shrub with biological controls aspatially. Cacho et al. (2006, 2008) presented a model in which actual spatial effort to find and eradicated invasive plant species was simulated using Monte Carlo methods with some spatial information and simulated decisions for when to attempt eradication. They found that search speed, eradication efficiency (actual kills for effort expended vs. failures for the same effort), and intrinsic biology (of seed longevity and germination rates) affected the optimization of effort. Cacho and Hester (2011) followed with an econometric approach analogous to production possibility frontiers. Rew et al. (2006) presented a model meant to optimize the allocation of effort in the detection of invasive species in Yellowstone National Park, USA, which was spatially explicit. They simulated a pattern of invasive species (cf. Seppelt 2005) and then sampled it in a GIS setting. They did not account for travel time to sites or for differences in time to cover the same distance but in different terrain.

A strategy implemented by CDRS was as follows (Atkinson, personal communication, June 2007):

> For blackberry on Floreana, the target is to find small seedlings and the surrounding vegetation is usually dense. A team of 5 people walks transects 5 m apart and can cover about 3 ha per day. If the vegetation is less dense the spacing can be increased to 10 m, but the rate of movement slows as one needs to scan a larger area per unit of forward movement. Mora adults are sprayed with herbicide and the seedlings pulled out by hand.
>
> For guava on Isabela, young plants (~2 year) are cut with a machete and the stem painted or sprayed with herbicide; larger individuals are girdled with a machete and the cut painted. This technique was successfully applied to Quinine (Buddenhagen et al. 2004) on Santa Cruz.

To possibly improve on these types of searches, we suggest that computer simulations of search behavior be used to compute efficiency. This approach borrows from models from location problems (e.g., Davydova and Romanovskii 1983) in optimizing supply chain management (Berman et al. 2011). Applying such a simulation in the Galapagos (and elsewhere) requires recognition of a number of constraints.

The important island-specific issue for ground searching is similar to that for remote sensing: the amount of time it takes to reach the starting point for a search pattern; this cost has been ignored in other studies (Rew et al. 2006). On any single island, such as Santa Cruz, the location of the Charles Darwin Research Station, there is very little road and the vast majority of the island can be reached only by horse—but only in the higher elevations—or foot across some combination of dense forest, dense spiny arid zones, and very rough footing. The island issue arises here because for many locations on a single island the best approach is to start by sea. For example, on Isabela, there is only one village with a Galapagos National Park (GNP) station on the southern edge of the island, and most of the islands, except for the slopes of Sierra Negra, are inaccessible from it. But the island issue is raised in getting from a regular base to the uninhabited islands. And access by sea presents its own challenges. First, there is time involved, and not just for the botanists/naturalists/rangers involved in the detection and eradication effort but also for either a GNP boat crew or a hired private boat. Both options add considerable expense, more so in the case of the private boat, but certain costs of using a GNP boat and crew, such as the opportunity cost of their not being on patrol for illegal fishing, are hard to calculate. Then, not everywhere is accessible even by boat. Much of the Galapagos coastline is wave-beaten cliffs where no landing is possible. Some islands have no safe landing at all. Others, where a safe landing can be made, have cliffs limiting any further access. All of these factors need to be computed in any optimization effort. The cost of travel time, computed as the wages of workers plus transport costs, to the site from the CDRS or GNP headquarters near Puerto Ayora would need to be added.

*Starting point.* Assuming that we know the type of environment in which we expect or fear invasives, we might pick a random point and from there begin a search pattern. More likely, one would begin based on how one reaches the environment, and so the origin will have spatial bias in being at the edge of the search zone.

*Width of survey.* How far one can see the invasives one is looking in a particular environment will determine the spacing of people walking parallel—if parallel is the choice—or how much area a person walking any line will cover. In most vegetation, sight distance is relatively low, and even in pampas grass where one might see 10 s of meters above the grass, small invasives below the grass level might not be seen at a distance of over 2 m.

*Walking speed.* In the Galapagos, relatively easy terrain in higher elevation pampas may be difficult to walk through because of the very dense grass at ankle level, while in the arid zone, a potentially impenetrable thicket dominated by prickly pear cactus (Opuntia spp.) is often underlain by the tortuous basalt of aa lava described by Darwin.

*Routes.* A common procedure for such a search is to run parallel transects across an area. This pattern can be varied by widening the gaps, given that the maximum width is uncertain, but then potentially missing an invader. An alternative search pattern is a spiral. Research has focused on optimization with spirals, both for general cases and applications in pattern recognition (e.g., Hall 1982; Tu et al. 2000), but how to best link multiple searchers will need to be solved. One can also envision a combination of transects and spirals, wherein a transect is walked until an individual invader is found and then a spiral out from there is walked and the transect renewed after some number of circuits. This combination may best approximate what is actually now done in the field by CDRS and GNP teams. Rew et al. (2006) even examined random walks, which can be useful as a null model but not necessary when existing procedures are being followed or modified. We could also develop an implementation of the branch-and-bound methodology that has been used to search parameter space in optimization problems (e.g., Davydova and Romanovskii 1983; Church and Gerrard 2003); this method involves adding short transects to a pixel wherein a target has been located but returning when additional targets are not found, and since it involves the retracing of steps, it may not be efficient in a real-space context unless retracing is less time consuming than moving along new routes.

For illustration purposes, we have developed a search pattern simulation using NetLogo. NetLogo is an easy to use yet powerful agent simulation package (Figs. 12.3a, b). In this simulation, we create a pattern of habitat into which we can place invasive individuals in varying degrees of spatial aggregation and adjust the search to meet the conditions described above. The actual distances and times would be recorded in the field and an average used for each type of vegetation. The time step can be recorded in NetLogo as clock ticks and the ticks tied to the environment. Into this environmental field, a search agent is placed, who then chooses a nearby starting point more likely to be in invasive habitat. Our implementation includes random walks, parallel transects, spirals, and transects plus spirals wherein an agent, upon encountering a cell with an invasive plant, walks a spiral before continuing on a transect. An upper bound on the amount of time available to search is set.

## Eradication

Eradication is more complex than it first appears (Renteria et al. 2012). Gardener et al. (2010) reviewed 30 invasive plant eradication projects on the Galapagos Islands. The most salient result is that eradication requires persistence. Single efforts are unsuccessful. In the optimization framework, the addition of eradication increases the time spent when detection is successful but also adds a dimension of repeated travel and effort where detection is not an issue.

Eradication would be added to an optimization model by determining the time needed to physically eradicate a plant and its effectiveness. Effectiveness must be used to schedule repeat visits, and current observations indicate that eradication is a multiyear endeavor because of seed banks and resprouting. The repeat visits skip the problems of detection, and the optimization question is tied to the value of the effort against relatively easily determined costs. The value question is discussed next.

## The Bigger Issue

Lastly and perhaps most problematically is the issue of weighting which places should get the most attention in detection and eradication. For optimization, the research frontier is the multi-scale tradeoff between allocating effort among islands—those already invaded or those still pristine—and then allocating effort in

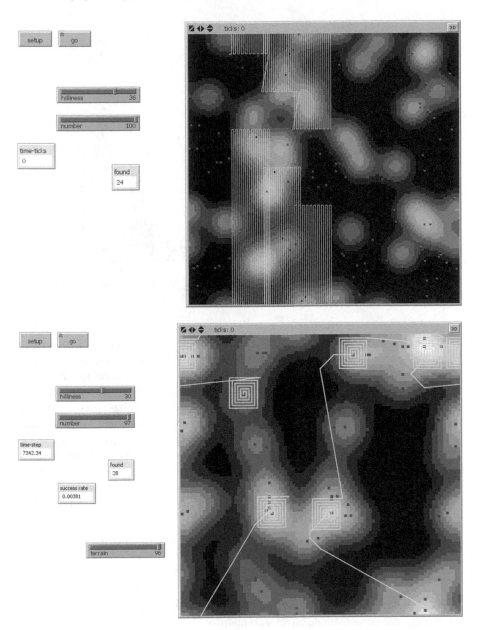

**Fig. 12.3** Two search routines in a NetLogo interface

different habitats and at what intensity within habitats. This problem applies to remote sensing in the cost of additional frames or flight lines and image processing as well as to fieldwork. In the Galapagos, the situation is stark: the larger inhabited islands with some local place to stay, ground vehicles available for use on the limited roads, and reasonably good connections by sea on small ferries are all sites

**Fig. 12.4** Fern-sedge zone on Isabela that has "nascent foci" (sensu Moody and Mack 1988) of guava invasives

of major ongoing invasions. The smaller inaccessible islands have fewer invasives and may be free of some of the most aggressive plant species such as guava and mora. Should attention be directed to limiting the further spread on the inhabited islands where the amount of area of invasives eradicated or the amount of area protected from new invasives is likely to be higher per hour or dollar spent, or should the pristine nature of the small, isolated islands somehow make their protection and maintenance as invasive-free worth a higher cost per unit area? Moody and Mack (1988) showed that attacking satellite populations was much more effective in reducing the overall spread of invasives than attacking core areas; however, they did not include the cost of finding the satellites. Many conservationists will argue that the pristine islands are worth "infinitely" more effort (Atkinson, personal communication). Two points seem uncontestable:

> We need to be able to quantify the costs of detection and eradication in the context of island biogeography and human geography; and
> The valuation of area on the different islands will be subjective and changeable, but the impacts of invasives in a variety of areas can and should be calculated in the future.

For eradication efforts, repeat visits have the advantage that locations are known, effort has some demonstrable effect, and additional detection searches can be added to the same trip to an area known to be at risk. This point argues in favor of work on the known invasions.

The current thinking on pristine areas could include such areas within islands undergoing invasion as well as wholly pristine islands. For example, an area of the treeless fern-sedge on Isabela that was a nascent foci for invasion by guava in 2007 (Fig. 12.4) should have had a high priority because it was known, the impact of an expansion of guava would be great (~complete replacement of a community with few remaining examples), and it was easy to reach (<1 h drive plus two on horseback). Beyond a research frontier, the effort needs to continue with the current approaches that are a result of experience, cost monitoring, and personal satisfaction, which should not be underestimated.

# References

Atkinson RJ, Rentería JL, Simbaña W (2008) The consequences of herbivore eradication on Santiago: are we in time to prevent ecosystem degradation again? In: Cayot L, Toral Granada V (eds) Galapagos Report 2007–2008. Charles Darwin Foundation, Galapagos National Park and INGALA, Puerto Ayora, Galapagos, pp 121–124

Bergstrom DM, Chown SL (1999) Life at the front: history, ecology and change on southern ocean islands. Trends Ecol Evol 14:472–477

Berman O, Krass D, Wang J (2011) The probabilistic gradual covering location problem on a network with discrete random demand weights. Comput Oper Res 38:1493–1500

Binggeli P, Hall JB, Healey JR (1998) A review of invasive woody plants in the tropics. School of Agricultural and Forest Sciences Publication 13. University of Wales, Bangor. http://www.safs.bangor.ac.uk/iwpt

Boardman JW (1993) Automated spectral unmixing of AVIRIS data using convex geometry concepts. Summaries 4th Jet Propulsion Laboratory, Airborne Geoscience Workshop, vol 1. Pasadena, California, pp 11–14.

Boardman JW, Kruse FA, Green RO (1995) Mapping target signatures via partial unmixing of AVIRIS data," in Proc. Summ. 5th Annu. JPL Airborne Earth Sci. Workshop, Pasadena, CA, pp. 23–26

Brand PR, Wiedenfeld DA, Snell HL (2012) Current status of alien vertebrates in the Galapagos Islands: invasion history, distribution, and potential impacts. Biol Invasions 14:461–480

Buddenhagen C, Jewell KJ (2006) Invasive plant seed viability after processing by some endemic Galapagos birds. Ornithologia Neotropical 17:73–80

Buddenhagen CE, Renteria JL, Gardener M, Wilkinson SR, Soria M, Yanez P, Tye A, Valle R (2004) The control of a highly invasive tree Cinchona pubescens in Galapagos. Weed Technol 18:1194–1202

Cacho OJ, Spring D, Pheloung P, Hester S (2006) Evaluating the feasibility of eradicating an invasion. Biol Invasions 8:903–917

Cacho OJ, Wise RM, Hester SM, Sinden JA (2008) Bioeconomic modeling for control of weeds in natural environments. Ecol Econ 65:559–568

Cacho OJ, Hester SM (2011) Deriving efficient frontiers for effort allocation in the management of invasive species. Aust J Agr Resource Econ 55:72–89

Carlqist SJ (1965) Island life. Natural History Press, New York

Carrion V, Donlan CJ, Campbell KJ, Lavoie C, Cruz F (2011) Archipelago-wide island restoration in the Galapagos Islands: reducing costs of invasive mammal eradication programs and reinvasion risk. PLOS One 6. e18835 DOI: 10.1371/journal.pone.0018835

Castillo JM, Leira-Doce P, Carrion-Tacuri J, Munoz-Guacho E, Arroyo-Solis A, Curado G, Doblas D, Rubio-Casal AE, Alvarez-Lopez AA, Redondo-Gomez S, Berjano R, Guerrero G, De Cires A, Figueroa E, Tye A (2007) Contrasting strategies to cope with drought by invasive and endemic species of Lantana in Galapagos. Biodiversity Conservation 16:2123–2136

Church RL, Gerrard RA (2003) The multi-level location set covering model. Geogr Anal 35:277–289

Coffey EED, Froyd CA, Willis KJ (2011) When is an invasive not an invasive? Macrofossil evidence of doubtful native plant species in the Galapagos Islands. Ecology 92:805–812

Cracknell AP, Hayes L (2007) Introduction to remote sensing, 2nd edn. Taylor and Francis, London

Cruz F, Carrion V, Campbell KJ, Lavoie C, Donlan CJ (2009) Bio-economics of large-scale eradication of feral goats from Santiago Island, Galapagos. J Wildl Manag 73:191–200

Davydova IM, Romanovskii IV (1983) The many-commodity location problem (the branch-and-bound method). Cybernetics 19:681–686

Dehaan R, Wilson LJ, Hall A, Rumbachs R (2007) Discrimination of blackberry (*Rubus fruticosus* sp. Agg.) using hyperspectral imagery in Kosciuszko National Park, NSW, Australia. J Photogramm Remote Sens 62:13–24

DeWalt SJ (2006) Population dynamics and potential for biological control of an exotic invasive shrub in Hawaiian rainforests. Biol Invasions 8:1145–1158

Dirnbock T, Greimler J, Lopez P, Stuessey TF (2003) Predicting future threats to the native vegetation of Robinson Crusoe Island, Juan Fernandez Archipelago, Chile. Conserv Biol 17:1650–1659

Eckhardt RC (1972) Introduced plants and animals in the Galapagos Islands. Bioscience 22:585–590

Ellshoff ZE, Gardner DE, Wikler C, Smith CW (1995) Annotated bibliography of the genus *Psidium*, with emphasis on *P. cattleianum* (strawberry guava) and *P. guajava* (common guava), forest weeds in Hawai'i. Cooperative National Park Resources Studies Unit, University of Hawai'i at Manoa. Tech Rep 95: pp 1–102

Everitt JH, Anderson GL, Escobar DE, Davis MR, Spencer NR, Andrascik RJ (1995) Use of remote sensing for detecting and mapping of leafy spurge (*Euphorbia esula*). Weed Technol 9:599–609

Gardener MR, Atkinson R, Renteria JL (2010) Eradications and people: lessons from the plant eradication program in Galapagos. Restor Ecol 18:20–29

Ge S, Carruthers R, Gong P, Herrera A (2006) Texture analysis for mapping *Tamarix parviflora* using aerial photographs along the Cache Creek, California. Environ Monit Assess 114:65–83

GISD (Global Invasive Species Database) (2005) Psidium guajava. http://www.issg.org/database/species/ecology.asp?si=211&fr=1&sts; last modified 8.16.10.

Gonzalez JA, Montes C, Rodriguez J, Tapia W (2008) Rethinking the Galapagos Islands as a complex social-ecological system: implications for conservation and management. Ecol Soc 13:Art.13.

Green AA, Berman M, Switzer P, Craig MD (1988) A transformation for ordering mutispectral data in terms of image quality with implications for noise removal. IEEE Trans Geosci Rem Sens 26:65–74

Guezou A, Pozo P, Buddenhagen C (2007) Preventing establishment: an inventory of introduced plants in Puerto Villamil, Isabela Island, Galapagos. PLOS One 2. e1042 DOI: 10.1371/journal.pone.0001042

Guezou A, Trueman M, Buddenhagen CE, Chamorro S, Guerrero AM, Pozo P, Atkinson R (2010) An extensive alien plant inventory from the inhabited areas of Galapagos. PLOS One 5. e10276 DOI: 10.1371/journal.pone.0010276

Hall RW (1982) Efficient spiral search in bounded spaces. IEEE Trans Pattern Anal Mach Intell 4:208–215

Hamman O (1979) Regeneration of vegetation on Santa Fe and Pinta Islands, Galapagos, after the eradication of goats. Biol Conserv 15:215–236

Hamman O (1981) Plant communities of the Galapagos Islands. Dansk Botanisk Arkiv 34(2):1–163

Hamann O (1984) Changes and threats to the vegetation. In: Perry R (ed) Key environments—Galapagos. Pergamon Press, Oxford, pp 115–132

Hamman O (1991) Indigenous and alien plants in the Galapagos Islands: problems of conservation and development. In: Heywood VH, Wyse Jackson PS (eds) Tropical botanic gardens: their role in conservation and development. Academic Press, London, pp 169–192

Hamann O (1993) On vegetation recovery, goats and giant tortoises on Pinta Island, Galapagos, Ecuador. Biodiversity Conserv 2:138–151

Hamann O (2004) Vegetation changes over three decades on Santa Fe Island, Galapagos, Ecuador. Nord J Bot 23:143–152

Heleno R, Blake S, Jaramillo P, Traveset A, Vargas P, Nogales M (2011) Frugivory and seed dispersal in the Galapagos: what is the state of the art? Integr Zool 6:110–129

Hobbs RJ (2001) Synergisms among habitat fragmentation, livestock grazing, and biotic invasions in southwestern Australia. Conserv Biol 15:1522–1528

Hobbs RJ, Mooney HA (2005) Invasive species in a changing world: the interactions between global change and invasives. In: Mooney HA, Mack RN, McNeely JA, Neville LE, Schei PJ, Waage JK (eds) Invasive alien species: a new synthesis. Island Press, Washington, DC, pp 310–331

Hunt ER Jr, Parker-Williams AE (2006) Detection of flowering leafy spurge with satellite multispectral imagery. Rangeland Ecol Manag 59:494–499

Itow S (2003) Zonation pattern, succession process and invasion by aliens in species-poor insular vegetation of the Galapagos Islands. Global Environ Res 7:39–58

Jäger H, Kowarik I (2010a) Resilience of native plant community following manual control of invasive *Cinchona pubescens* in Galapagos. Restor Ecol 18:103–112

Jäger H, Kowarik I, Tye A (2009) Destruction without extinction: long-term impacts of an invasive tree species on Galapagos highland vegetation. J Ecol 97:1252–1263

Jäger H, Tye A, Kowarik I (2007) Tree invasion in naturally treeless environments: Impacts of quinine (Cinchona pubescens) trees on native vegetation in Galapagos. Biol Conserv 140:297–307

Jäger H, Kowarik I (2010b) Resilience of native plant community following manual control of invasive *Cinchona pubescens* in Galapagos. Restor Ecol 18(Suppl 1):103–112

Kupfer JA, Malanson GP, Franklin SB (2006) Not seeing the ocean for the islands: the mediating influence of matrix-based processes on forest fragmentation effects. Global Ecol Biogeogr 15:8–20

Lass LW, Prather TS, Glenn NF, Weber KT, Mundt JT, Pettingill J (2005) A review of remote sensing of invasive weeds and examples of the early detection of spotted knapweed (*Centaurea maculosa*) and babysbreath (*Gypsophila paniculata*) with a hyperspectral sensor. Weed Sci 53:242–251

Lavoie C, Donlan CJ, Campbell K, Cruz F, Carrion GV (2007) Geographic tools for eradication programs of insular non-native mammals. Biol Invasions 9:139–148

Levine JM, Vila M, D'Antonio CM, Dukes JS, Grigulis K, Lavorel S (2003) Mechanisms underlying the impacts of exotic plant invasions. Proc Roy Soc Lond B 270:775–781

Loope LL, Mueller-Dombois D (1989) Characteristics of invaded islands with special reference to Hawaii. In: Drake JA, Mooney HA, di Castri F, Groves RH, Kruger FJ, Rejmanek M, Williamson M (eds) Biological Invasions, a Global Perspective. SCOPE 37. Wiley, New York, pp 257–280.

MacArthur RH, Wilson EO (1967) The theory of Island biogeography. Princeton University Press, Princeton, NJ

Macdonald IAW, Ortiz L, Lawesson JE, Nowak JB (1988) The invasion of highlands in Galapagos by the red quinine tree *Cinchona succirubra*. Environ Conservat 15:215–220

MacFarland C, Cifuentes M (1996) Case study: Galapagos, Ecuador. In: Dompka V (ed) Human population, biodiversity and protected areas: science and policy issues. American Association for the Advancement of Science, Washington, DC, pp 135–188

Mack RN, Simberloff D, Lonsdale WM, Evans H, Clout M, Bazzaz FA (2000) Biotic invasions: causes, epidemiology, global consequences, and control. Ecol Appl 10:689–710

Maheu-Giroux M, de Blois S (2005) Mapping the invasive species *Phragmites australis* in linear wetland corridors. Aquat Bot 83:310–320

Matisziw TC, Murray AT (2006) Promoting species persistence through spatial association optimization in nature reserve design. J Geogr Syst 8:289–305

Mauchamp A (1997) Threats from alien plant species in the Galapagos islands. Conserv Biol 11:260–263

Mauchamp A, Aldaz I, Ortiz E, Valdebenito H (1998) Threatened species, a re-evaluation of the status of eight endemic plants of the Galapagos. Biodiversity Conserv 7:97–107

Miao X, Gong P, Swope SM, Pu R, Carruthers RI, Anderson GL (2006) Estimation of yellow starthistle abundance through CASI-2 hyperspectral imagery using linear spectral mixture models. Rem Sens Environ 101(3):329–341

Miller BW, Breckheimer I, McCleary AL, Guzman-Ramirez L, Caplow SC, Jones-Smith JC, Walsh SJ (2010) Using stylized agent-based models for population-environment research: a case study from the Galapagos islands. Popul Environ 31:401–426

Mireya Guerrero A, Tye A (2011) Native and introduced birds of Galapagos as dispersers of native and introduced plants. Ornitologia Neotropical 22:207–217

Moody ME, Mack RN (1988) Controlling the spread of plant invasions: the importance of nascent foci. J Appl Ecol 25:1009–1021

Mooney HA, Cleland EE (2001) The evolutionary impact of invasive species. Proc Nat Acad Sci U S A 98:5446–5451

NRC Committee on the Scientific Basis for Predicting the Invasive Potential of Nonindigenous Plants and Plant Pests in the United States (2002) Predicting invasions of nonindigenous plants and plant pests. National Academy Press, Washington, DC

Mooney HA, Hobbs RJ (eds) (2000) Invasive species in a changing world. Island Press, Washington, DC

Mundt JT, Glenn NF, Weber KT, Pather TS, Lass LW, Pettingill J (2005) Discrimination of hoary cress and determination of its detection limits via hyperspectral image processing and accuracy assessment techniques. Rem Sens Environ 96:509–517

Naylor BJ, Endress BA, Parks CG (2005) Multiscale detection of sulfur cinquefoil using aerial photography. Rangeland Ecol Mgmt 58:447–451

Parker IM, Simberloff D, Lonsdale WM, Goodell K, Wonham M, Kareiva PM, Williamson MH, Von Holle B, Moyle PB, Byers JE, Goldwasser L (1999) Impact: toward a framework for understanding the ecological effects of invaders. Biol Invasions 1:3–19

Porter DM (1976) Geography and dispersal of Galapagos islands vascular plants. Nature 264:745–746

Porter DM (1979) Endemism and evolution in Galapagos islands vascular plants. In: Bramwell D (ed) Plants and islands. Academic Press, London, pp 225–256

Regan TJ, McCarthy MA, Baxter PWJ, Panetta FD, Possingham HP (2006) Optimal eradication: when to stop looking for an invasive plant. Ecol Lett 9:759–766

Rew LJ, Maxwell BD, Aspinall R (2005) Predicting the occurrence of nonindigenous species using environmental and remotely sensed data. Weed Sci 53:236–241

Rew LJ, Maxwell BD, Dougher FL, Aspinall R (2006) Searching for a needle in a haystack: evaluating survey methods for non-indigenous species. Biol Invasions 8:523–539

Rentería JL, Buddenhagen C (2006) Invasive plants in the *Scalesia pedunculata* forest at Los Gemelos, Santa Cruz, Galapagos. Galapagos Res 64:31–35

Renteria JL, Atkinson R, Buddenhagen C (2007) Estrategias para la erradicacion de 21 especies de plantas. Fundacion Charles Darwin, Puerto Ayora, Galapagos, Ecuador

Renteria JL, Gardener MR, Panetta FD, Crawley MJ (2012) Management of the invasive hill raspberry (*Rubus niveus*) on Santiago Island, Galapagos: eradication or indefinite control? Invasive Plant Sci Manag 5:37–46

Ricciardi A (2007) Are modern biological invasions an unprecedented form of global change? Conserv Biol 21:329–336

Richardson DM, Pysek P, Rejmanek M, Barbour MG, Panetta FD, West CJ (2000) Naturalization and invasion of alien plants: concepts and definitions. Diversity Distrib 6:93–107

Schofield EK (1989) Effects of introduced plants and animals on island vegetation: examples from the Galapagos Archipelago. Conserv Biol 3:227–238

Seppelt R (2005) Simulating invasions in fragmented habitats: theoretical considerations, a simple example and some general implications. Ecol Complexity. 2:219–231

Sharma GP, Singh JS, Raghubanshi AS (2005) Plant invasions: emerging trends and future implications. Curr Sci 88:726–734

Simberloff D, Von Holle B (1999) Positive interactions of nonindigenous species: invasional meltdown? Biol Invasions 1:21–32

Space JC, Falanruw M (1999) Observations on invasive plant species in Micronesia. http://www.hear.org/AlienSpeciesInHawaii/articles/pier/pier_micronesia_report.pdf. Accessed 18 Apr 2012.

Space JC, Flynn T (2000) Observations on invasive plant species in American Samoa. http://www.hear.org/alienspeciesinhawaii/articles/pier/pier_samoa_report.pdf. Accessed 18 Apr 2012.

St Quinton JM, Fay MF, Ingrouille M, Faull J (2011) Characterization of *Rubus niveus*: a prerequisite to its biological control in oceanic islands. Biocontrol Sci Technol 21:733–752

Stockwell CA, Hendry AP, Kinnison MT (2003) Contemporary evolution meets conservation biology. Trends Ecol Evol 18:94–101

Trueman M, Atkinson R, Guezou A, Wurm P (2010) Residence time and human-mediated propagule pressure at work in the alien flora of Galapagos. Biol Invasions 12:3949–3960

Tu YM, Li B, Niu JW (2000) A novel motion estimation algorithm based on dynamic search window and spiral search. Lect Notes Comput Sci 1948:356–362

Tye A (2006) Can we infer island introduction and naturalization rates from inventory data? Evidence from introduced plants in Galapagos. Biol Invasions 8:201–215

Tye A (2001) Invasive plant problems and requirements for weed risk assessment in the Galapagos islands. In: Groves RH, Panetta FD, Virtue JG (eds) Weed risk assessment. CSIRO Publishing, Collingwood, Victoria Australia, pp 153–175

Tye A (2002) Revision of the threat status of the endemic flora of Galapagos. Galapagos Report 2001–2002. World Wildlife Fund—Fundación Natura, Quito, pp 116–122

Tye A, Francisco-Ortega J (2011) Origins and evolution of Galapagos endemic vascular plants. In: Bramwell D, Caujapé-Castells J (eds) The biology of island floras. Cambridge University Press, London, pp 89–153

Underwood E, Ustin SL, DiPierto D (2003) Mapping non-native plants using hyperspectral imagery. Rem Sens Environ 86:150–161

Walsh SJ, McCleary AL, Mena CF, Shao Y, Tuttle JP, Gonzalez A, Atkinson R (2008) QuickBird and Hyperion data analysis of an invasive plant species in the Galapagos islands of Ecuador: implications for control and land use management. Rem Sens Environ 112:1927–1941

Watson J, Trueman M, Tufet M, Henderson S, Atkinson R (2010) Mapping terrestrial anthropogenic degradation on the inhabited islands of the Galapagos Archipelago. Oryx 44:79–82

White PCL, Harris S (2002) Economic and environmental costs of alien vertebrate species in Britain. In: Pimentel D (ed) Biological invasions—economic and environmental costs of alien plant, animal, and microbe species. CRC Press, Boca Raton, FL, pp 113–149

Wilkinson SR, Naeth MA, Schmiegelow FKA (2005) Tropical forest restoration within Galapagos National Park: application of a state-transition model. Ecol Soc 10:Art. 28.

Williams AP, Hunt ER (2002) Estimation of leafy spurge cover from hyperspectral imagery using mixture tuned matched filtering. Rem Sens Environ 82:446–456

Wiser SK, Allen RB, Clinton PW, Platt KH (1998) Community structure and forest invasion by an exotic herb over 23 years. Ecology 79:2071–2081

# Chapter 13
# From Whaling to Whale Watching: Cetacean Presence and Species Diversity in the Galapagos Marine Reserve

**Judith Denkinger, Javier Oña, Daniela Alarcón, Godfrey Merlen, Sandy Salazar, and Daniel M. Palacios**

## Introduction

When sperm whales became scarce in the Atlantic Ocean in the eighteenth century, whalers ventured for new whaling grounds in the Pacific, and in 1792 Captain James Colnett from the British whaling fleet described Galapagos as teeming with whales. Sperm whales (*Physeter macrocephalus*) in their Galapagos breeding grounds were no longer safe (Hickman 1985). After 1812, more than 700 whalers from the US alone and others from Norway, Britain, and Peru removed approximately 5,000 animals from the islands (Hope and Whitehead 1991) and dramatically reduced local populations of Galapagos fur seals (*Arctocephalus galapagoensis*) and tortoises (*Chelonidis elephantopus*) as well.

In the twentieth century, the world began to see the Galapagos as a priceless treasure for wildlife, and since 1930, it began to put in force a series of decrees leading to the creation of the Galapagos National Park in 1959 and the Galapagos Marine Reserve (GMR) in 1986. Today, all commercially hunted whales are either threatened or close to extinction and are officially protected under the International Convention for the Trade of Endangered Species (CITES), the International Union

J. Denkinger (✉) • J. Oña • D. Alarcón
Galapagos Science Center, College of Biological and Environmental Sciences,
Universidad San Francisco de Quito, Quito, Ecuador
e-mail: jdenkinger@usfq.edu.ec

G. Merlen
Galapagos National Park, Puerto Ayora, Santa Cruz, Ecuador

S. Salazar • D.M. Palacios
Joint Institute for Marine and Atmospheric Research, University of Hawaii,
1000 Pope Road, Marine Sciences Building, Room 312, Honolulu, HI 96822, USA

NOAA/NMFS/SWFSC/Environmental Research Division,
1352 Lighthouse Avenue, Pacific Grove, CA 93950-2097, USA

S.J. Walsh and C.F. Mena (eds.), *Science and Conservation in the Galapagos Islands:* 217
*Frameworks & Perspectives*, Social and Ecological Interactions in the Galapagos Islands 1,
DOI 10.1007/978-1-4614-5794-7_13, © Springer Science+Business Media, LLC 2013

for the Conservation of Nature (IUCN), the International Whaling Commission (IWC), and Ecuadorian legislations, such as the Whale Sanctuary created in 1990 and covering not only waters around Galapagos but extending to the entire 200 nautical mile exclusive economic zone (Merlen 1992; Hoyt 2005).

In recent years, the waters around the Galapagos have been identified as one of the focal areas for marine mammals in the eastern tropical Pacific (Ferguson et al. 2003) and are considered one of the global hot spots for marine mammal diversity (MacLeod and Mitchell 2006; Pyenson 2011; Kaschner et al. 2011), with at least 23 species of cetaceans recorded inside the GMR (Day 1994; Palacios and Salazar 2002).

In the past two decades, whale watching has become increasingly popular in Latin America, increasing at an average rate of 11.3% from 1995 to 2008 (Hoyt and Iñíguez 2008), with Galapagos emerging as one of the hot spots for wildlife tourism. However, new concerns have arisen regarding the effects of climate change on the distribution of cetaceans (see Whitehead et al. 2008; MacLeod 2009; Simmonds and Isaac 2007; Kaschner et al. 2011), as warming ocean temperatures are predicted to lead to reduced species diversity in tropical and subtropical environments (Whitehead et al. 2008; Gambaiani et al. 2009). Galapagos is situated within the area of direct influence of the El Niño Southern Oscillation (ENSO) phenomenon and with every El Niño event the marine environment suffers drastic changes caused by excessive heating of surface waters, nutrient stress, and food shortages that lead to the reduction of higher trophic level predator populations such as pinnipeds (Trillmich and Limberger 1985), penguins (Boersma 1998; Vargas et al. 2006), and flightless cormorants (Valle and Coulter 1989). In contrast to these resident predators, cetaceans can move and avoid food shortages, but there is little information about cetacean migrations in and out the GMR (Palacios et al. 2010). Wade and Gerrodette (1993) report a considerable interannual variability of species abundance and presence during their line transects surveys in the eastern tropical Pacific (ETP) from 1986 to 1990. In this context, cetacean presence in the Galapagos can help to understand the current situation of whales and dolphins under changing oceanographic conditions. Thus, long-term data sets are important to assess changes in species composition and possible species loss in the GMR.

Using wildlife tourism in the Galapagos as a research platform, we discuss the situation of cetaceans in the GMR over an 18-year period covering El Niño, La Niña, and neutral conditions, using long-term data sets of occasional sightings reported by trained tour guides as well as from dedicated research cruises. We also provide a description of the general patterns of occurrence for cetacean species and discuss some of the most common species in detail.

## Study Area

The Galapagos Archipelago, with 13 large islands, is situated 100 km west of continental South America, where it extends from 3° north to 4° south latitude and 87–94° west longitude. Our study area was restricted to regular tourist navigation routes in the center and south of the GMR.

The GMR is characterized by changing oceanographic conditions due its proximity to the equatorial front in the north of the archipelago (Palacios 2004). It is influenced by two major ocean currents—the South Equatorial and the Equatorial Undercurrent or Cromwell Current (Fiedler and Talley 2006). From the north, the Panama Bight brings warm surface waters with average sea surface temperatures (SST) of 27°C (Palacios 2004) and causes a warm, less productive season from December to May. From June to November, the Humboldt or Peru Current from the south is more prevalent with strong winds and cold, productive waters with average SST of 22°C. The Cromwell Current from the west flows at approximately 100 m depth and collides with the Galapagos Islands in the west off of Fernandina and Isabela islands, producing strong upwelling plumes that extend to the south and north–central portion of the archipelago (Palacios and Salazar 2002). Galapagos is situated in the center of the "ENSO Region 3,4" (Sweet et al. 2007), where El Niño events cause drastic declines in productivity, especially from December to February until La Niña events respond with negative anomalies in SST and high productivity (see Guilyardi et al. 2009).

## Methods

### Data Collection and Surveys

Species presence in the GMR was established using published records and direct observations from 1993 to 2010 by trained tour guides from Lindblad Expeditions on the MS *Endeavor* and MS *Islander* (Fig. 13.1). For each sighting, information was recorded including date, time, position, and area of the sighting as well as species, number of animals observed, and general behavior. A total of 1,407 sightings were analyzed for presence, dominance, and occurrence for species with at least 20 sightings in total.

Species presence was analyzed as absence/presence records for each year from 1993 to 2010. The number of species sighted each year was compared during El Niño, La Niña, and neutral conditions using ENSO patterns provided by the US National Oceanic and Atmospheric Administration. Common species were selected when they were sighted at least 20 times throughout the study period. Seasonal preference of the most common species was evaluated using T-tests with 95% confidence intervals during warm seasons (December to May) and cold seasons (June to November), adapted from Palacios (2003) to facilitate evaluation of baleen whale presence during Northern Hemisphere and Southern Hemisphere summer months.

### Photo ID Studies on Orca

To determine orca movements in the GMR, orcas were photographically identified using natural marks on the dorsal fin (Hammond et al. 1990; Würsig and

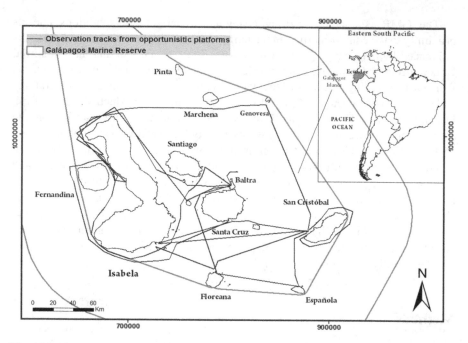

**Fig. 13.1** Location of the Galapagos Archipelago and the Galapagos Marine Reserve with the research area according to observation tracks from opportunistic platforms

Jefferson 1990). Only good quality photographs were used for comparison and pods were determined by the identification of different group members sighted at the same time and location.

## Results and Discussion

### *Cetacean Species Diversity in the Galapagos Marine Reserve*

A total of 26 cetacean species in six families are reported in the database we compiled for the GMR (Table 13.1). In the present study, 23 species were recorded as occasional sightings during tourist and research cruises. Overall, 11 species are considered rare and were only seen one or two times, while 12 species are considered common, with more than 20 sighting recorded over the study period. The most common species, such as bottlenose dolphins (*Tursiops truncatus*), Bryde's whales (*Balaenoptera edeni*), common dolphins (*Delphinus delphis*), orcas (*Orcinus orca*) and humpback whales (*Megaptera novaeangliae*) make up 70% of the recorded sightings. Among the baleen whales, Bryde's whales are frequently seen, whereas minke whales (*Balaenoptera acutorostrata*), blue whales (*B. musculus*), and fin

**Table 13.1**  Cetaceans reported for the Galapagos Marine Reserve

| | Species name | Type of observation | Source |
|---|---|---|---|
| *Balaenopteridae* | | | |
| Minke whale | *Balaenoptera acutorostrata* | Live sighting | GSC/CDR/GNP |
| Sei whale | *Balaenoptera borealis* | Live sighting | GSC/CDR/GNP |
| Bryde's whale | *Balaenoptera edeni* | Live sighting | GSC/CDR/GNP |
| Blue whale | *Balaenoptera musculus* | Live sighting | GSC/CDR/GNP |
| Fin whale | *Balaenoptera physalus* | Live sighting | GSC/CDR/GNP |
| Humpback whale | *Megaptera novaeangliae* | Live sighting | GSC/CDR/GNP |
| *Delphinidae* | | | |
| Common dolphin | *Delphinus delphis* | Live sighting | GSC/CDR/GNP |
| Pigmy killer whale | *Feresa attenuata* | Live sighting | CDR/GNP |
| Short-finned pilot whale | *Globicephala marcorhynchus* | Live sighting | GSC/CDR/GNP |
| Risso's dolphin | *Grampus griseus* | Live sighting | GSC/CDR/GNP |
| Fraser dolphin | *Lagenodelphis hosei* | Live sighting | CDR/GNP |
| Killer whale | *Orcinus orca* | Live sighting | GSC/CDR/GNP |
| Melon-headed dolphin | *Peponocephala electra* | Live sighting | (Merlen 1995; Smith (1999); Palacios and Salazar 2002) |
| False killer whale | *Pseudorca crassidens* | Live sighting | (Merlen 1995; Palacios et al. 2004) |
| Pantropical spotted dolphin | *Stenella attenuata* | Live sighting | (Palacios and Salazar 2002; Palacios 2003) |
| Striped dolphin | *Stenella coeruleoalba* | Live sighting | (Palacios 2003; Palacios 1999) |
| Spinner dolphin | *Stenella longirostris* | Live sighting | GSC |
| Rough toothed dolphin | *Steno bredanensis* | Live sighting | (Merlen 1995; Palacios and Salazar 2002; Palacios et al. 2004) |
| Bottlenose dolphin | *Tursiops truncatus* | Live sighting | (Palacios 1999) |
| *Kogiidae* | | | |
| Pigmy sperm whale | *Kogia sima* | Live sighting | GSC/CDR/GNP |
| *Physeteridae* | | | |
| Sperm whale | *Physeter macrocephalus* | Live sighting | GSC/CDR/GNP |
| *Ziphiidae* | | | |
| Ginkgo-toothed beaked whale | *Mesoplodon ginkgodens* | Stranding record | (Palacios et al. 2004) |
| Pygmy-beaked whale | *Mesoplodon peruvianus* | Live sighting | Daniel Palacios/ocean alliance, unpublished |
| Cuvier's beaked whale | *Ziphius cavirostris* | Live sighting | GSC/CDR/GNP |
| Longman beaked whale | *Indopacetus pacificus* | Live sighting | (Pitman et al. 1999; Palacios and Salazar 2002) |
| Blainville's beaked whale | *Mesoplodon densirostris* | Live sighting | (Pitman et al. 1999; Palacios and Salazar 2002) |

*GSC* Galapagos Science Center, *CDR* Charles Darwin Research Station, *GNP* Galapagos National Park Service

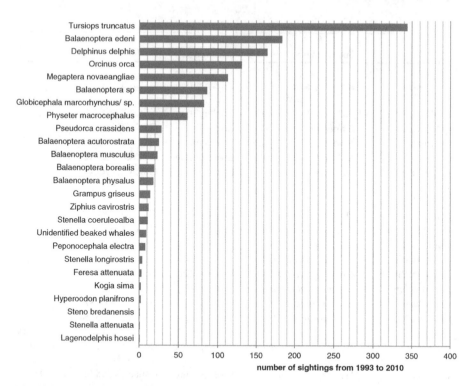

**Fig. 13.2** Cumulative sighting records for cetacean species in the Galapagos Marine Reserve from observations collected by tour naturalist guides from 1993 to 2010

whales (*B. physalus*) are considered to be rare in the GMR (Fig. 13.2). While Palacios and Salazar (2002) rank common dolphins as the second most common species, in this study, Bryde's whales were seen far more frequently. This is due, however, to our focus on waters near the islands, within the regular tourist routes, while Palacios and Salazar (2002) and Palacios (2003) based their results on dedicated cetacean surveys that included more pelagic areas of the GMR.

Overall, species numbers fluctuate over the years with a decreasing amount of species being reported during strong El Niño years, such as 1997/1998 and 2010. Despite cold, productive conditions provided by a strong La Niña event in 1999, most species were not sighted until 2001. During the cooler conditions from 2001 to 2008, species numbers peaked from 2001 to 2003, but decreased with the onset of the moderate 2004 El Niño event. Sightings increased again during the cooler La Niña conditions during 2007 and 2008, but diminished to a total of six species during the stronger El Niño event in 2010 (Fig. 13.3).

Most of the species observed were already registered as common in previous surveys, conducted from 1985 to 1995 (Palacios 1999; Palacios and Salazar 2002), but Risso's dolphins (*Grampus griseus*), described as common, are now only very occasionally sighted, whereas dolphins of the genus *Stenella*, Frasers dolphins (*Lagenodelphis hosei*), remain very rare and generally absent during El Niño years, although during offshore research cruises in the eastern tropical Pacific from 1986

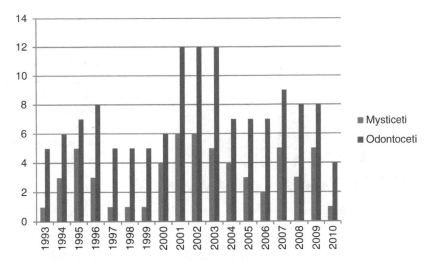

**Fig. 13.3** Presence and absence analysis of mysticete and odontocete species in the Galapagos Marine Reserve from 1993 to 2010 using the total number of species sighted each year. ENSO years with moderate El Niño; strong El Niño; moderate La Niña, strong La Niña years (adapted from http://www.cpc.ncep.noaa.gov)

to 1990 spinner, common, and striped dolphins were most common (Wade and Gerrodette 1993). The lack of sightings of Risso's dolphins may be a concern, since they share the same habitat with the frequently seen bottlenose dolphins (Palacios 2003), but the first recorded sightings in our study appeared in 2002—they are only seen during cooler periods and La Niña conditions, though more dedicated surveys are needed to confirm the current distribution of this species. Nonetheless, most species seem to move off the GMR during El Niño years, which confirms recent findings on shifting geographic ranges to remain in a certain niche (Lambert et al. 2011; Kaschner et al. 2011; Whitehead et al. 2008).

All species were observed throughout the year, but only humpback and blue whales showed a significant preference toward the colder, second semester from June to November. This would coincide with the migration of Southeast Pacific whales. According to observations by Merlen (1995) and Smith (1999), most odontocetes, except for the common dolphin, false killer whale and sperm whale, and all other baleen whales, are more frequently seen during cooler seasons, although no significant seasonality could be observed in this study (Table 13.3).

## Distribution

Most of the baleen and toothed whales were sighted west off Isabela Island in the Bolivar Channel between Isabela and Fernandina islands (Figs. 13.4 and 13.5). Baleen whales seem to have a clear preference for the cooler and more productive waters of the south and west of the GMR. Most of the blue whales were seen

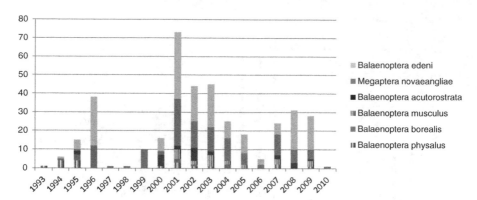

**Fig. 13.4** Sighting frequency of baleen whales per total baleen whale sightings from 1993 to 2010

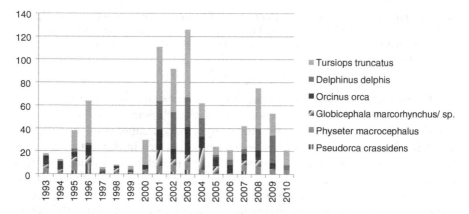

**Fig. 13.5** Sighting frequencies of the most common odontocetes per total odontocete sightings from 1993 to 2010

in the south and west of Floreana and Isabela islands. Humpback and Bryde's whales are also frequently sighted in the warmer and less productive central archipelago and east to San Cristobal Island, while minke whales venture into the more tropical, productive area of the south of Genovesa (Fig. 13.6). Odontocetes are present throughout the GMR, even north to Darwin and Wolf islands. Sightings in the more productive southwest, however, are more frequent. Orcas move between all the islands with numerous sightings in the Bolivar Channel, within Santiago, Baltra and San Cristobal islands, and north of Floreana Island (Figs. 13.7 and 13.9). Sperm whales seem to prefer the deep waters of the south and west of Isabela Island, but some sperm whales were seen in shallower waters south of Marchena Island (Fig. 13.7).

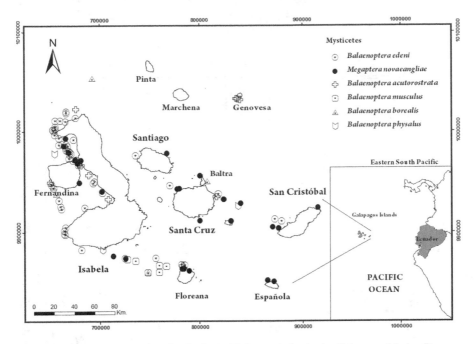

**Fig. 13.6** Cumulative sighting distribution of baleen whales in the Galapagos Marine Reserve from 1993 to 2010

## Case Studies

### Bottlenose Dolphins (*Tursiops truncatus*)

According to the numerous sightings throughout the years, bottlenose dolphins seem to be resident in the GMR. Overall, sightings increased during cooler years, which coincide with the fact that bottlenose dolphins prefer coastal upwelling systems (Palacios 2003), where most of the observations during cruises are made, which may explain the huge proportion of dolphin sightings. Bottlenose dolphins are curious animals and they frequently "bow ride" with cruise boats, which also increases sighting probability.

### Common Dolphins (*Delphinus delphis*)

Common dolphins were reported from 1995 to the onset of the 1997 El Niño; disappeared during 1998 and 2000; and returned from 2001 to the end of the study period, including 2010, but were relatively rare from 2004 to 2007. Though they are one of the most common species in the GMR, according to Smith (1999) and Palacios and Salazar (2002), sightings associated with this study became less frequent, especially during warmer

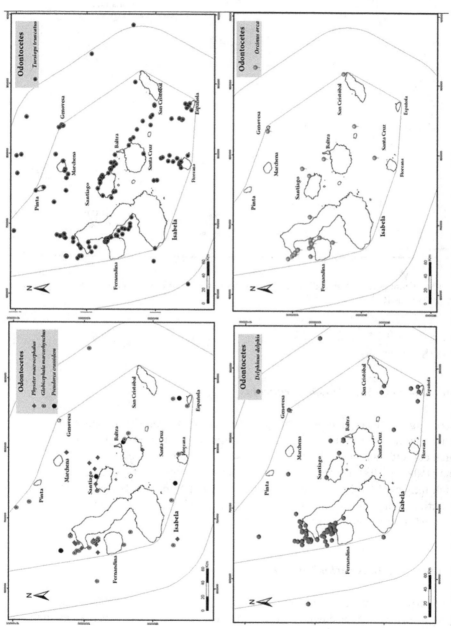

**Fig. 13.7** Cumulative sighting distribution of odontocetes in the Galapagos Marine Reserve from 1993 to 2010

years, as no common dolphins were seen during El Niño years; possibly supporting their emigration to other geographic ranges as suggested by Lambert et al. (2011) for common dolphins off the coast of California, USA (Table 13.2 and Fig. 13.5).

**Orcas (*Orcinus orca*)**

Orcas were present during all years and all seasons of the study with a slightly higher number of recorded sightings during the cold season (Tables 13.2, 13.3 and Fig. 13.5). Most of the orcas were seen in the Bolivar Channel, but occasionally they appeared at Punta Vicente Roca, Fernandina Island, North Seymour Island, San Cristobal Island at Punta Pitt and León Dormido, and Floreana Island at Punta Cormoran (See Fig. 13.7d).

A total of 17 orcas were photographically identified (Fig. 13.8), with a total of 27 sightings and seven resighted animals (see Table 13.4). The relatively high number of resightings suggests a small orca population for Galapagos of less than the estimated orca abundance of 156 animals (Alava 2009), with recorded sighting. The 17 orcas identified for Galapagos can be grouped into four different pods, with an average pod size of 3.2 animals/pod (Table 13.4), consistent with Merlen (1999) who defined an average pod size of 3.1 animals, using sighting recorded from 1948 to 1997. The largest pod, "Pod No. 2" is formed by seven animals, while the other groups observed were of two, four, and five animals, respectively. Four orcas appeared to be solitary animals. The relatively small pod size of Galapagos orcas is characteristic and comparable to transient orcas off the coast of British Columbia, Canada, who feed on marine mammals among other prey (Baird and Dill 1996). Orca attacks on sea lions, dolphins, and whales have been observed in Galapagos (e.g., Brennan and Rodríguez 1994; Merlen 1999), and on humpback whales off the coast of Manabi and Esmeraldas on mainland Ecuador (Denkinger personal observation). Three of the animals were resighted within several years with the longest resighting interval of 6 years from 2005 to 2011, which suggests the presence of a small resident orca population in the GMR.

**Sperm Whales (*Physeter macrocephalus*)**

Even though Galapagos became a famous hot spot for whaling in the eighteenth century, sperm whales today only account for about 6% of all the sightings (Fig. 13.1). Though the absence of sperm whale sightings in 1997 and 1999 through 2006 seems to be independent of El Niño/La Niña events, as well as the neutral years, the number of sightings fluctuated throughout the years, but never exceeded eight sightings per year (Table 13.2 and Fig. 13.5). Sperm whales in the Galapagos have been decreasing at a rate of 20% per year in recent years, as the low calving rates are unsustainable even though feeding success seems to be comparable to

**Table 13.2** Cetacean species presence/absence in the Galapagos Marine Reserve from 1993 to 2010

| Species/years | 1993 | 1994 | 1995 | 1996 | 1997 | 1998 | 1999 | 2000 | 2001 | 2002 | 2003 | 2004 | 2005 | 2006 | 2007 | 2008 | 2009 | 2010 |
|---|---|---|---|---|---|---|---|---|---|---|---|---|---|---|---|---|---|---|
| Balaenopteridae | | | | | | | | | | | | | | | | | | |
| *B. acutorostrata* | 0 | 0 | 1 | 0 | 0 | 0 | 0 | 1 | 1 | 1 | 1 | 0 | 0 | 0 | 1 | 1 | 1 | 0 |
| *B. borealis* | 0 | 0 | 1 | 1 | 0 | 0 | 0 | 0 | 1 | 1 | 0 | 1 | 0 | 0 | 0 | 0 | 0 | 0 |
| *B. edeni* | 0 | 1 | 1 | 1 | 0 | 0 | 0 | 1 | 1 | 1 | 1 | 1 | 1 | 1 | 0 | 1 | 1 | 0 |
| *B. musculus* | 0 | 0 | 0 | 0 | 0 | 0 | 0 | 0 | 1 | 1 | 1 | 1 | 1 | 0 | 1 | 0 | 1 | 0 |
| *B. physalus* | 1 | 1 | 1 | 0 | 0 | 0 | 0 | 1 | 1 | 1 | 1 | 0 | 0 | 0 | 0 | 0 | 0 | 0 |
| *Megaptera novaeangliae* | 0 | 1 | 1 | 1 | 1 | 1 | 1 | 1 | 1 | 1 | 1 | 1 | 1 | 1 | 1 | 1 | 1 | 1 |
| Delphinidae | | | | | | | | | | | | | | | | | | |
| *Delphinus delphis* | 0 | 0 | 1 | 1 | 1 | 0 | 1 | 0 | 1 | 1 | 1 | 1 | 1 | 1 | 1 | 1 | 1 | 1 |
| *Feresa attenuata* | 0 | 0 | 0 | 1 | 0 | 0 | 0 | 1 | 0 | 1 | 0 | 0 | 0 | 0 | 0 | 0 | 0 | 0 |
| *G. marcorhynchus* | 1 | 1 | 1 | 1 | 0 | 1 | 1 | 1 | 0 | 1 | 1 | 1 | 0 | 1 | 1 | 0 | 1 | 0 |
| *Grampus griseus* | 0 | 1 | 0 | 1 | 0 | 0 | 0 | 1 | 1 | 0 | 1 | 0 | 0 | 1 | 0 | 0 | 0 | 0 |
| *Lagenodelphis hosei* | 0 | 0 | 0 | 0 | 0 | 0 | 0 | 0 | 0 | 0 | 0 | 0 | 0 | 0 | 0 | 0 | 0 | 0 |
| *Orcinus orca* | 1 | 1 | 1 | 1 | 1 | 1 | 1 | 1 | 1 | 1 | 0 | 1 | 1 | 1 | 1 | 1 | 1 | 1 |
| *Peponocephala electra* | 0 | 0 | 0 | 0 | 0 | 0 | 0 | 1 | 0 | 1 | 0 | 0 | 0 | 1 | 0 | 0 | 0 | 0 |
| *Pseudorca crassidens* | 1 | 1 | 1 | 1 | 1 | 1 | 1 | 1 | 0 | 1 | 1 | 1 | 1 | 0 | 0 | 0 | 0 | 0 |
| *Stenella attenuata* | 0 | 0 | 0 | 0 | 0 | 0 | 0 | 0 | 0 | 1 | 1 | 0 | 0 | 0 | 1 | 0 | 0 | 0 |
| *Stenella coeruleoalba* | 0 | 0 | 0 | 0 | 0 | 0 | 0 | 0 | 0 | 0 | 0 | 0 | 0 | 1 | 1 | 1 | 1 | 0 |
| *Stenella longirostris* | 0 | 1 | 0 | 0 | 0 | 0 | 0 | 0 | 0 | 0 | 1 | 0 | 0 | 0 | 0 | 0 | 0 | 0 |
| *Steno bredanensis* | 0 | 0 | 0 | 0 | 0 | 0 | 0 | 0 | 0 | 0 | 0 | 0 | 0 | 0 | 0 | 0 | 0 | 1 |
| *Tursiops truncatus* | 1 | 1 | 1 | 1 | 1 | 1 | 1 | 1 | 1 | 1 | 1 | 1 | 1 | 1 | 1 | 1 | 1 | 1 |
| Physeteridae | | | | | | | | | | | | | | | | | | |
| *P. macrocephalus* | 1 | 1 | 1 | 1 | 0 | 1 | 0 | 1 | 1 | 1 | 0 | 1 | 0 | 0 | 1 | 1 | 1 | 1 |
| Kogiidae | | | | | | | | | | | | | | | | | | |
| *Kogia sima* | 0 | 0 | 0 | 0 | 0 | 0 | 0 | 0 | 0 | 1 | 0 | 0 | 0 | 0 | 1 | 0 | 0 | 0 |
| Ziphiidae | | | | | | | | | | | | | | | | | | |
| *Ziphius cavirostris* | 0 | 0 | 0 | 0 | 0 | 0 | 0 | 0 | 1 | 1 | 1 | 1 | 1 | 0 | 1 | 1 | 1 | 0 |
| *Indopacetus pacificus* | 0 | 0 | 0 | 0 | 0 | 0 | 0 | 0 | 1 | 0 | 0 | 0 | 0 | 0 | 0 | 0 | 0 | 0 |

**Table 13.3** Seasonality according to the cumulative number of sightings during warm and cold season of the 12 most common species observed in the Galapagos Marine Reserve

| Species | Warm season | Cold season | P value[a] |
|---|---|---|---|
| *Tursiops truncatus* | 172 | 194 | >0.05 |
| *Orcinus orca* | 61 | 70 | >0.05 |
| *Delphinus delphis* | 92 | 73 | >0.05 |
| *Globicephala macrorhynchus* | 36 | 46 | >0.05 |
| *Pseudorca crassidens* | 16 | 12 | >0.05 |
| *Physeter macrocephalus* | 33 | 28 | >0.05 |
| *Balaenoptera acutorostrata* | 10 | 16 | >0.05 |
| *Balaenoptera borealis* | 8 | 11 | >0.05 |
| *Balaenoptera edeni* | 89 | 92 | >0.05 |
| *Balaenoptera musculus* | 3 | 20 | 0.029 |
| *Balaenoptera physalus* | 6 | 12 | >0.05 |
| *Megaptera novaeangliae* | 18 | 93 | 0.013 |

[a]$T$-test significance with 95% CI

**Fig. 13.8** Identified Orcas in the Galapagos Marine Reserve through April 2012

other more successful populations (Whitehead et al. 1997). A sighting of a group with 50 sperm whales off Isabela Island in February 2008 (Jonathan Aguas, personal communication) is encouraging; however, most of the sightings reported here consisted of single animals or small groups with less than 10 animals.

**Table 13.4** Sighting records of photo identified orcas and pods in the Galapagos Marine Reserve

| ID number | Pod number and composition (as of April 2012) | Sex | Sightings and range |
|---|---|---|---|
| Oo003 | 1 (2 Adult males) | Male | October 2007–North Seymour |
| Oo004 | 1 (2 Adult males) | Male | October 2007–North Seymour |
| Oo005 | No ID | Male | September 2008–Punta Pitt (Cristobal Island) |
| Oo006 | 2 (2 Adult males, 1 juvenile male, 4 adult females) | Female | May 2005–Bolivar Channel; December 2011–Punta Cormoran (Floreana); February 2012–Bolivar Channel and Kicker Rock (San Cristobal); March 2012–Bolivar Channel |
| Oo007 | 2 (2 Adult males, 1 juvenile male, 4 adult females) | Female | May 2011–Bolivar Channel; December 2011–Punta Cormoran (Floreana) |
| Oo008 | No ID | No ID | December 2011–Punta Cormoran (Floreana) |
| Oo009 | 3 (5 Animals, 1 adult male, 2 females, 1 juvenile, 1 calf) | Calf | September 2008–Punta Pitt (San Cristobal) |
| Oo010 | 2 (2 Adult males, 1 juvenile male, 4 adult females) | Juvenile Male | February 2012–Kicker Rock (San Cristobal), March 2012–Bolivar Channel; April 2012–Bolivar Channel |
| Oo011 | Single | Male | 2005–Fernandina; May 2011–Punta Vicente Roca; September 2011–Bolivar Channel |
| Oo012 | 2 (2 Adult males, 1 juvenile male, 4 adult females) | Male | February 2012–Bolivar Channel, March 2012–Bolivar Channel |
| Oo013 | 2 (2 Adult males, 1 juvenile male, 4 adult females) | Female | February 2012–Bolivar Channel |
| Oo014 | 2 (2 Adult males, 1 juvenile male, 4 adult females) | Female | February 2012–Bolivar Channel and Kicker Rock (San Cristobal); |
| Oo015 | 2 (2 Adult males, 1 juvenile male, 4 adult females) | Juvenile Male | February 2012–Bolivar Channel; March 2012–Bolivar Channel |
| Oo016 | 2 (2 Adult males, 1 juvenile male, 4 adult females) | Juvenile | March 2012–Bolivar Channel |
| Oo017 | 2 (7 Animals) | Juvenile | March 2012–Bolivar Channel |
| Oo018 | 4 (1 Female, 3 juveniles) | Female | September 2010–Punta Vicente Roca |
| Oo019 | Single | | July 2011–Punta Vicente Roca |

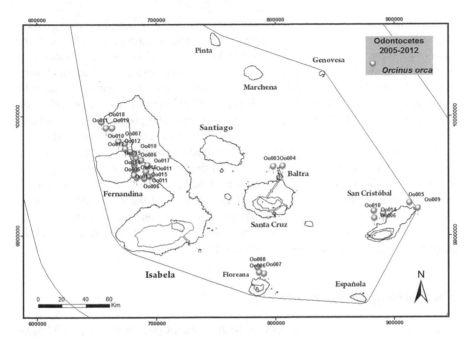

**Fig. 13.9** Distribution of identified orca sightings in the Galapagos Marine Reserve from 2005 to April 2012

## Bryde's Whales (*Balaenoptera edeni*)

Bryde's whales seem to have disappeared during the 1997/1998 El Niño event and were not seen until 2000. From 2000 to 2009 they were present during all years, but disappeared again in 2010 during the last strong El Niño event (Table 13.2, Fig. 13.4). This species seems to avoid waters of low productivity, as it is more common during the cold season (Table 13.4). Bryde's whales feed in Galapagos waters, where they mostly occur in the very productive upwelling areas in the western portions of the archipelago (Palacios and Salazar 2002; Palacios 2003; Alava 2009). Bryde's whales, however, are also frequently seen north and west of San Cristobal Island throughout the year, where even mother-calf pairs were observed (Denkinger, personal observation).

## Humpback Whales (*Megaptera novaeangliae*)

Interestingly, humpback whales, even though less frequently sighted than Bryde's whales, were seen every year since 1994, including El Niño years and are now a common species for the GMR (Table 13.2 and Fig. 13.4). Just as blue whales, humpback whales show a clear preference for the cold season that coincides with the Southern Hemisphere winter migration of the Southeast Pacific population, to which the humpback whales at the Ecuadorian breeding grounds also belong (Alava and Felix 2006; Félix et al. 2011). While the abundance of humpback whales off the coast of Ecuador peaks in July and August (Alava et al. 2011), most of the sightings in the GMR are at the end of the breeding season in September/October, and hump-back whales are seen year round. Humpback whales were seen feeding off Santa Fe Island and south of Isabela Island in February (Denkinger, personal observation), suggesting that the GMR may support a regularly migrating segment of the popula-tion, but more photo ID studies are necessary to support this hypothesis. Galapagos has been considered a breeding area for humpback whales (Félix et al. 2011), but the fact that most whales are observed late in the breeding season, most sightings are of single animals or mother-calf pairs, and the lack of competitive male groups could also suggest that Galapagos is used as a stepping stone for whales migrating to the Panamanian/Colombian breeding grounds. This contention is supported by Félix et al. (2011), who found genetic homogeneity of one biopsied female in the Galapagos with haplotypes from Colombian humpback whales.

**Blue Whales (*Balaenoptera musculus*)**

Blue whales are an endangered species with a current population of possibly less than 1% of the natural population, because of whaling, including catches in Galapagos (Branch et al. 2007). In the GMR, they were considered to be rare (Palacios 1999), but, since 2001, the number of blue whale sightings has increased and they have been consistently present, with the exception of 2006 and 2010 (Table 13.2 and Fig. 13.4). Galapagos blue whales have a significant preference for the colder season from June to November that coincides with migrations of the Southern Chile population to low-latitude waters (Branch et al. 2007). Blue whale sightings from whaling vessels off the coast Peru from 1976 to 1983 peaked during the southern summer from January to March (Reilly and Thayer 1990), which is the opposite of our observations. Three sightings, however, were made during the warmer season with one sighting in February of a group of three feeding blue whales at Caleta Iguana, along southern Isabela Island (Denkinger, personal observation). This may indicate that some animals remain in the Galapagos Islands throughout the year or that there is a possible exchange of Northern Hemisphere blue whales wintering near the Costa Rica Dome.

# Conclusions

Cetacean presence in the GMR clearly decreases during El Niño years and increases during cooler La Niña years. Overall, there were fewer species reported during the last 10 years, which could signal a shift in species composition in the GMR. But as the sighting data were opportunistically collected, and only reflect the tourist routes, animal sightings data are not directly comparable with studies based on systematic surveys that include larger areas of the GMR.

Common species such as Bryde's, humpback, and blue whales are sighted throughout the year, with only humpback whales and blue whales showing a clear preference for the cooler, second semester of the year. This indicates that some baleen whales may be resident to the GMR.

All odontocetes are present throughout the year, especially in the more productive waters west of Isabela Island, with bottlenose dolphins being the most common species in the GMR. Orcas are frequently sighted in the Bolivar Channel, but so far four pods have been identified in the region located west and south-central of the GMR. Of the 17 photographically identified orcas, eight were resighted several times and at different locations, which suggests a small resident orca population and an overall small number of orcas in the GMR. Overall, occasional sighting reports are an important tool to detect long-term changes in the composition of the cetacean community.

Although there is great interest in whale watching, this tourism activity has not yet fully developed in Galapagos. The results of this study provide information about the species and the areas where tourists visiting the GMR might expect to see cetaceans more regularly. It is clear that Galapagos supports a unique and diverse cetacean fauna that can be reliably observed along the established routes for tourism vessels, and this information could be the basis for the establishment of a targeted whale watching industry. These operations, nonetheless, should take into account the conservation status and particular responses of the different species to natural environmental fluctuations like ENSO in order to implement an organized and responsible activity.

**Acknowledgments**   We are especially grateful to the tour guides of Lindblad Expeditions and others for providing us with numerous sighting records and to the Galapagos National Park for taking us along and hosting us at Bolivar Channel. Our thanks go to Chiara Guidino, Annika Krutwa, and all the other volunteers; to Walter Traunspurger and Hubert Spieth from Bielefeld University for their support on a Southwestern Galapagos Research cruise; and to Ben Haase for sharing his Orca photographs. Finally, we express our appreciation to Pol Segarra and Carlos Mena for editing sighting positions, and to Eduardo Espinoza for helpful comments. The present project has been carried out under Research Permit (PC-23-10 and PC-27-11) of the Galapagos National Park.

# References

Alava JJ (2009) Carbon productivity and flux in the marine ecosystems of the Galapagos Marine Reserve based on cetacean abundances and trophic indices. Revista de Biologia Marina y Oceanografía 44(1):109–122

Alava JJ, Felix F (2006) Logistic population curves and vital rates of the southeastern pacific humpback whale stock off Ecuador. Report of the IWC Workshop on Comprehensive Assessment of Southern Hemisphere Humpback Whales, Hobart, Tasmania, 3–7 Apr 2006. SC-A-06HW1

Alava JJ, Barragán MJ, Denkinger J (2011) Assessing the impact of bycatch on Ecuadorian humpback whale breeding stock: a review with management recommendations. Ocean Coast Manag 57:34–43

Baird RW, Dill LM (1996) Ecological and social determinants of group size in transient killer whales. Behav Ecol 7(4):408–416

Boersma PD (1998) Population trends of the Galapagos penguin: impacts of El Niño and La Niña. Condor 100:245–253

Branch TA, Stafford KM, Palacios DM, Allison C, Bannister JL, Burton CLK, Cabrera E et al (2007) Past and present distribution, densities and movements of blue whales *Balaenoptera musculus* in the Southern Hemisphere and northern Indian Ocean. Mamm Rev 37(2):116–175

Brennan B, Rodríguez P (1994) Report of two orca attacks on cetaceans in Galapagos. Noticias de Galapagos 54:28–29

Day D (1994) List of cetaceans seen in Galapagos. Noticias de Galapagos 53:5–6

Félix F, Palacios DM, Salazar SK, Caballero S, Haase B, Falconí J (2011) The 2005 Galapagos humpback whale expedition: a first attempt to assess and characterize the population in the archipelago. J Cetacean Res Manag (Special Issue) 3:291–299

Ferguson MC, Barlow J, Shores LJ (2003) Addendum: spatial distribution and density of cetaceans in the eastern tropical pacific ocean based on summer/fall research vessel surveys in 1986–1996 National Marine Fisheries Service, NOAA, La Jolla, California 92038. Administrative Report LJ-01-04

Fiedler PC, Talley L (2006) Hydrography of the eastern tropical Pacific: a review. Progr Oceanogr 69:143–180

Gambaiani DD, Mayo P, Isaac SJ, Simmonds MP (2009) Potential impact of climate change and greenhouse gas emissions on Mediterranean marine ecosystems and cetaceans. J Mar Biol Assoc UK 89:179–201

Guilyardi E, Wittenberg A, Fedorov A, Collins M, Wangm C, Capotondi A, van Oldenborgh GJ (2009) Understanding El Niño in ocean–atmosphere general circulation models: progress and challenges. Bull Am Meteorol Soc 90(3):325

Hammond PS, Mizroch SA, Donovan P (1990) Individual recognition of cetaceans: use of photo-identification and other techniques to estimate population parameters. Rep Int Whaling Comm (Special Issue 12):143–145

Hickman J (1985) The enchanted islands: the Galapagos discovered. Noticias de Galapagos 42:26–27

Hope PL, Whitehead H (1991) Sperm whales off the Galapagos Islands from 1830–50 and comparisons with modern studies. Rep Int Whaling Comm (Special Issue) 12:135–139

Hoyt E (2005) Marine protected areas for whales, dolphins and porpoises: a worldwide handbook for cetacean habitat conservation. Earthscan, London, 516 pp

Hoyt E, Iñíguez M (2008) The state of whale watching in Latin America. WDCS, IFAW, Chippenham, UK; Yarmouth Port, USA; and Global Ocean, London

Kaschner K, Tittensor DP, Ready J, Gerrodette T, Worm B (2011) Current and future patterns of global marine mammal diversity. PLoS One 6(5):e19653

Lambert E, MacLeod C, Hall K, Brereton T, Dunn T, Wall D, Jepson P (2011) Quantifying likely cetacean range shifts in response to global climatic change: implications for conservation strategies in a changing world. Endangered Species Res 15(3):205–222

MacLeod CD, Mitchell G (2006) Key areas for beaked whales worldwide. J Cetacean Res Manag 7(3):309–322

MacLeod CD (2009) Global climate change, range changes and potential implications for the conservation of marine cetaceans: a review and synthesis. Endangered Species Res 7:125–136

Merlen G (1992) Ecuadorian whale refuge. Noticias de Galapagos 51:23–24

Merlen G (1995) A field guide to the marine mammals of Galapagos. Instituto Nacional de Pesca, Guayaquil, Ecuador

Merlen G (1999) The orca in Galapagos: 135 sightings. Noticias de Galapagos. 60–63

Palacios DM (1999) Blue whale (*Balaenoptera musculus*) occurrence off the Galapagos Islands, 1978–1995. J Cetacean Res Manag 1(1):41–51

Palacios DM (2003) Oceanographic conditions around the Galapagos Archipelago and their influence on cetacean community structure. Dissertation, Oregon State University

Palacios DM (2004) Seasonal patterns of sea-surface temperature and ocean color around the Galapagos: regional and local influences. Deep-Sea Res II 51(1–3):43–57

Palacios DM, Salazar SK (2002) Cetáceos. In: Danulat E, Edgar GJ (eds) Reserva Marina de Galapagos. Línea Base de la Biodiversidad. Fundación Charles Darwin/Servicio Parque Nacional Galapagos, Santa Cruz, Galapagos, Ecuador

Palacios DM, Salazar SK, Day D (2004) Cetacean remains and strandings in the Galapagos Islands, 1923–2003. Lat Am J Aquat Mamm 3(2):127–150

Palacios DM, Salazar SK, Vargas FH (2010) Galapagos marine vertebrates: responses to environmental variability and potential impacts of climate change. In: Larrea I, Di Carlo G (eds) Climate change vulnerability assessment of the Galapagos Islands. World Wildlife Fund and Conservation International, USA, pp 69–80

Pyenson ND (2011) The high fidelity of the cetacean stranding record: insights into measuring diversity by integrating taphonomy and macroecology. Proc Royal Soc B: Biol Sci 278(1724):3608–3616

Pitman RL, Palacios PM, Brennan PLR, Brennan BJ, Balcomb KC, Miyashita T (1999) Sightings and possible identity of a bottlenose whale in the tropical Indo-Pacific: *Indopacetus pacificus*? Mar Mamm Sci 15(2):531–549

Simmonds MP, Isaac S (2007) The impacts of climate change on marine mammals: early signs of significant problems. Oryx 41:19–26

Smith SD (1999) Distribution of dolphins in Galapagos waters. Mar Mamm Sci 15(2):550–555

Sweet WV, Morrison JM, Kamykowski D, Schaeffer BM, Banks S, McCulloch A (2007) Water mass seasonal variability in the Galapagos Archipelago. Deep-Sea Res I 54:2023–2035

Trillmich F, Limberger D (1985) Drastic effects of El Niño on Galapagos pinnipeds. Oecologia 67(1):19–22

Valle CA, Coulter MC (1989) Present status of the flightless cormorant, Galapagos penguin and greater flamingo populations in the Galapagos Islands, Ecuador after the 1982–82 El Niño. Condor 89:276–281

Vargas FH, Harrison S, Rea S, Macdonald DW (2006) Biological effects of El Niño on the Galapagos penguin. Biol Conserv 127(1):107–114

Wade PR, Gerrodette T (1993) Estimates of cetacean abundance and distribution in the Eastern Tropical Pacific. Rep Int Whaling Comm 43:477–494

Whitehead HS, Cristal J, Dufault S (1997) Past and distant whaling and the rapid decline of sperm whales off the Galapagos Islands. Conserv Biol 11(6):1387–1396

Whitehead H, MacGill B, Worm B (2008) Diversity of deep-water cetaceans in relation to temperature: implications for ocean warming. Ecol Lett 11:1198–1207

Würsig B, Jefferson TA (1990) Methods of photo-identification for small cetaceans. Rep Int Whaling Comm (Special Issue 12). 43–52

# Index

**A**
ABMs. *See* Agent-based models (ABMs)
Accommodation, 35, 59, 118
Agent-based models (ABMs)
    characteristics, 56
    climate change, 54, 61–62
    DEM, 57
    demographic system, 53
    economic development, 55
    exogenous factors, 53, 56
    external factors, 56
    feedback loops, 53
    Landscape Module, 57
    land use/land cover pattern, 62–63
    livelihoods, 58
    non-agent environment, 53
    organization, 54
    "pushes and pull" factors, 54
    resource conservation, 55
    socioeconomic and ecological
        subsystems, 53
    spatial simulation model, 54
    system behaviors and dynamics, 55
    tourism system
        *GalaSim* model, 60
        market operators and tourists, 60
        multi-scale patterns, 59
        overnight tours, 59
        selective tourism, 59
    virtual model, 54–55
Agricultural and urban areas
    bird diversity, 191
    ENSO, 188
    exotic species, 187
    fieldwork, 190
    human activities, 187
    natural ecosystems, 187
    organic agriculture site (*see* Organic
        agriculture site)
    pature and guava site (*see* Pature
        and guava site)
    quantitative analyses, 191
    restoration site (*see* Restoration site)
    soil invertebrates, 190
    terrestrial ecosystems, 188
    urban site (*see* Urban site)
    vegetation cover, 190
Analysis of variance (ANOVA), 191
Araucaria Project, 39

**B**
*Bacteroides spp.*, 179–181
Baleen whales, 225
Biological conservation, 11
Blackberry, 203
Blue-footed boobies, 179
Blue Whales *(Balaenoptera musculus)*,
    222, 234
Bottlenose dolphins *(Tursiops truncatus)*,
    222, 227
Bryde's whales *(Balaenoptera edeni)*,
    222, 233

**C**
Carbon concentration
    organic agriculture site, 192, 195
    pature and guava site, 192, 195
    restoration site, 192, 195
    urban site, 192, 195
CASI-2 hyper-spectral imagery, 206
Charles Darwin Foundation (CDF), 28, 38–39,
    95, 97, 133, 144

Charles Darwin Research Station (CDRS),
    14, 15, 46, 88, 99, 202
Childbirth, 149
Childcare, 81
Child health, 149–151
*Cinchona succirubra*, 203
Climate change, 41
    agent-based models, 54, 61–62
    glacial and interglacial cycles, 5
    island ecosystems, 198
Climate research, 4–5
Coastal water quality
    fecal contamination, Santa Cruz and San
        Cristobal islands
        education and benefits, 183–184
        improved water infrastructure, 183
        monitoring, 182–183
        study area and sample analysis,
            179–181
        study results and implications, 181–182
    water resources
        impacts, 177–178
        impairment, 176–177
        infrastructure, 177
*Coffea cf. arabica*, 188
Common Dolphins *(Delphinus delphis)*,
    227, 228
Complexity theory, 52–53. *See also*
    Agent-based models (ABMs)
Crabs, 178
Cruise boat tourism, 30


**D**
*Delphinus delphis*, 222
Demographic analysis
    birth rate, 72, 74
    distribution, 71
    "habitual" residents, 72
    immigration, 71, 73
    marriage and divorce statistics, 72, 73
    migrants, 72
    population structure, 71
    sex and age structure, 74
    tourists, 72, 73
Digital Elevation Model (DEM), 57


**E**
Ecological crisis
    CDRS, 99
    Charles Darwin Research Station, 95–96
    fishing cooperatives, 94
    FUNDAR, 96
    GNPS, 94

"Human Footprint" program, 99–100
    invasive plant species, control of, 98–99
    invasive species eradication, 97–98
    park management plan, 95
    policies and programs, 96
    rural environmentality, 100–102
    UNESCO designation, 94
Ecology research, 5
Ecotourism, 91
    boat and *cupo*, 122, 123
    culture of conservation, 132
    economy, 132
    environmental disasters, 123–124
    fund, government institutions, 135–137
    human services, 121
    immigration, 132
    institutions, 132–133
    model policy, 140–141
    pests and diseases, 123
    policy environment, 133–134
    1998 Special Law
        local leaders, 138–140
        Marine Reserve, 138
        subsidies, 137
    sport fishing, 122
    2010 Sustainable Tourism Summit, 124
Ecuador. *See* Galapagos
Elephant grass *(Pennisetum purpureum)*, 204
El Niño Southern Oscillation (ENSO),
    56, 188, 220
Employment, 54, 75, 121
Encyclopedia of Life (EOL), 87
*Enterococcus spp.*, 179–181
Environmentalism, 39
Environmental management practices, 94
*Escherichia coli (E. coli)*, 176
Evolution
    biology research, 5–6
    divergent selection, 9
    ecological opportunity, 9–11
    founder effect, 8
    gene flow, 9
    genetic drift, 8
    linear, 7
    multiple isolation, 9


**F**
Fecal contamination
    education and benefits, 183–184
    improved water infrastructure, 183
    monitoring, 182–183
    study area and sample analysis, 179–181
    study results and implications, 181–182
Fernandina Island, 3, 225, 229

Ferns, 188
Fishing industry, 32, 133, 134
Flightless cormorant *(Phalacrocorax harrisi)*, 7
Floating hotels, 109
Food and nutrition, 151
Foundation for Alternative Responsible Development (FUNDAR), 90, 96
Founder effects, 8
Frasers dolphins *(Lagenodelphis hosei)*, 224

**G**
Galapagos Islands
    affinity anomaly, 25
    agent-based models (*see* Agent-based models (ABMs))
    agricultural and urban areas (*see* Agricultural and urban areas)
    awareness, 41
    biogeographical reasons, 25
    bridging concepts, 37
    *carapachudos*, 31
    coastal water quality (*see* Coastal water quality)
    conceptual strategy, 93
    conservation and conservationism, 33
    Darwinism, 26
    demographic analysis (*see* Demographic analysis)
    ecological crisis (*see* Ecological crisis)
    economic transformations, 37–38
    ecotourism industry, 91
    Ecuadorian public's view, 30
    Eibl-Eibesfeldt's idea, 28
    environmentalism, 39
    environmental management practices, 94
    evolution (*see* Evolution)
    fauna and flora, 28
    fishermen and conservationists, 39–40
    fishing activities, 32–33
    governmental and nongovernmental organizations, 39
    health situation analysis (*see* Public health)
    history, 131
    human-environment relations, 93
    imperfect complex forms, 26
    invasive species (*see* Invasive species)
    isolation and connectivity, 27
        biophysical and socioeconomic factors, 42
        local culture, 43
        products and goods, 43
        standard of living, 43
        threats, 42
    labor market, 36
    lobsters and fish, 32
    mocking bird species, 25
    natural hazards, 93
    organism arrival, 6
    organism establishment, 6–7
    *pesca vicencial*, 39
    population growth, 91, 144
    replacement anomaly, 25
    sea lions, 38
    socioeconomic analysis (*see* Socioeconomic analysis)
    socio-environmental conflicts, 91
    Special Law, 92
    techniques and methods, 27
    tourism (*see* Tourism)
    UNESCO, 37
Galapagos Marine Reserve (GMR), 31
    GNP, 16
    1998 Special Law, 138
    whales
        cetacean species diversity, 222–225
        data collection and surveys, 221
        distribution, 225–227
        photo ID studies, 221–222
        wildlife tourism, 220
Galapagos National Institute (INGALA), 133
Galapagos National Park (GNP), 16, 51, 73, 89, 90, 133, 135, 136
Galapagos National Park Service (GNPS), 94, 107
Galapagos penguin *(Spheniscus mendiculus)*, 7
Galapagos sea lion *(Zalophus wollebaeki)*, 2
*Galapagueño*, 107, 111
Gene flow, 9
Genetic drift, 8
Genovesa Island, 3, 226
Geological research, 2–3
*Geospiza sp.*, 191
Gray-level co-occurrence matrix (GLCM), 160
Guava, 172, 203, 204

**H**
Health care, 81
Healthcare, Isla Isabela
    childbirth, 149
    child health and medical care, 149–151
    depression, 153
    medical infrastructure, 148–149
    overweight and obesity, 152–153
Health prevention coverage, 81
History, 131
"Hot spot"-specific approach, 183

"Human Footprint" program, 99–100
Humpback whales *(Megaptera
      novaeangliae)*, 222, 233–234
Hybrid culture. *See* Galapagos Islands


**I**
Immigration
   demographic analysis, 71, 73
   ecotourism, 132
   Special Law, 140
Industrial boats, 32
Institute of Forests and Natural Areas
      (INEFAN), 136
Intergovernmental Panel on Climate Change
      (IPCC), 188
International Convention for the Trade of
      Endangered Species (CITES), 219
International Union for the Conservation
      of Nature (IUCN), 219–220
International Whaling Commission
      (IWC), 220
Invasive species
   blackberry, 203
   elephant grass *(Pennisetum
         purpureum)*, 204
   eradication, 202, 209
   guava, 203, 204
   "nascent foci," Isabela, 212
   plant research, 204–205
   quinine, 204
   remote sensing
      aircraft platforms, 207
      AVIRIS imagery, 206
      field surveys, 208–210
      Hyperion imagery, 206
      image classification, 207
      LIDAR systems, 207
      Quickbird, 206
      vegetation indices, 207
      WorldView-2, 206
   spatial optimization, 205–206
Isabela Islands
   agriculture, 171
   data analysis, 146
   data collection methods, 145–146
   economic opportunities, 147
   food and nutrition, 151
   healthcare
      childbirth, 149
      child health and medical care, 149–151
      depression, 153
      medical infrastructure, 148–149
      overweight and obesity, 152–153

   health problems, 147–148
   land use and land cover information
      in 2004-2010, 166, 167
      image interpretation, 158
      remote sensing, 158
      Santo Tomás *(see* Santo Tomás)
   open and friendly atmosphere, 147
   safe environment, 147
   treeless fern-sedge, "nascent foci,"
   water and sanitation issues, 152
Island biocomplexity, 63, 64


**L**
Lagoons, 178
Land-based tourism
   advertising strategy, 36
   ecotourism packages, 36
   local residents and politicians, 35
   national (Ecuadorian) tourism, 35
   tourists, accommodation, 35
Landscape Module, 57
*Lantana camara*, 203
*Lecocarpus darwinii*, 188


**M**
Macroinvertebrate diversity
   organic agriculture site, 196
   pature and guava site, 196
   restoration site, 196
   urban site, 196
Marchena Island, 3, 226
Marine ecosystems, 13–14
Marine iguana *(Amblyrhynchus cristatus)*, 2
Marine Reserve, 138
Marine wildlife, 179
Medical care, 149–151
Minimum Noise Fraction (MNF), 206
Minke whales *(Balaenoptera
      acutorostrata)*, 222
Mixture-Tuned Matched Filter (MTMF)
      method, 206
Mollusk, 178


**N**
National park, 143
NetLogo interface, 211
Nitrogen concentration
   organic agriculture site, 192, 194
   pature and guava site, 192, 194
   restoration site, 192, 194
   urban site, 192, 194

**O**

Object-based image analyses (OBIA), 63
Oceanic fauna, 13
Oceanographic research, 4–5
Off-farm employment, 171
*On the Origin of Species*, 131
Optimization model. *See* Invasive species
Orcas *(Orcinus orca)*, 222, 229
Organic agriculture site
    carbon concentration, 192, 195
    location, 189
    macroinvertebrate diversity, 196
    nitrogen concentration, 192, 194
    variables, 193
    vegetation cover, 195
Overnight boat tour model, 59


**P**

Pature and guava site
    carbon concentration, 192, 195
    location, 189
    macroinvertebrate diversity, 196
    nitrogen concentration, 192, 194
    variables, 193
    vegetation cover, 195
*Pennisetum purpureum*, 203, 204
Phytoplankton, 178
Pinta Island, 3, 15
*Psidium guajava*, 188, 203
Public health
    beds availability, establishment, 83
    childcare, institution, 82
    health prevention coverage, 81
    morbidity, establishment, 83
    point of care, birth assistance, 84
    social security coverage, 81
    vaccine, 81, 82
Public wastewater systems, 183


**Q**

QuickBird satellite image, 160
Quinine, 204


**R**

Remote sensing
    aircraft platforms, 207
    AVIRIS imagery, 206
    field surveys, 208–210
    Hyperion imagery, 206
    image classification, 207
    LIDAR systems, 207
    Quickbird, 206

    vegetation indices, 207
    WorldView-2, 206
Residents
    crime and household waste, 118
    cultue of value, 132
    culture of conservation, 132
    economy class boats, 30
    "habitual" residents, 72
    native, 138
    permanent, 138
    socioeconomic conditions, 77
    Special Law, 117
    tourism, 122
Restoration site
    carbon concentration, 192, 195
    location, 189
    macroinvertebrate diversity, 196
    nitrogen concentration, 192, 194
    variables, 193
    vegetation cover, 195
Risso's dolphins *(Grampus griseus)*, 224, 225
Rose apple, 169
*Rubus niveus*, 203
Rural environmentality, 100–102


**S**

*Salmonella enterica* infection, 179
San Cristobal island, 30–32, 143, 175.
        *See also* Fecal contamination
Sanitation, 152
Santa Cruz Island
    fecal contamination (*see* Fecal
        contamination)
    fishermen and conservationists, 33
    population distribution, 71
    rural environmentality, 100
Santiago Island, 12, 15
Santo Tomás
    agriculture, 166
    barren land, 169
    climate, 159
    elevation ranges, 159
    field data and classification scheme, 160–161
    forest/shrub cover, 167, 169
    national park, 159
    object-based image analysis approach
        Definiens Professional , 5,162
        QuickBird image, 163–165
        segmentation parameters, 162
        WorldView-2 data, 162
    satellite image data and preprocessing, 160
    sociodemographic data and analysis,
        164, 166, 169–171
    vegetation, 159

Scalesia forests, 204
*Scalesia pedunculata*, 190
Scientific research, 1–2
SCUBA Iguana, 39
Sea cucumber *(Stichopus fuscus)*, 13
Sea lions, 7, 38, 178
Sewage. *See* Fecal contamination
Sewage treatment plants, 176
Social capital, 139
Social security coverage, 81
Socioeconomic analysis
    favorable conditions
        "economically active population," ,
            75,76
        information and communication
            technologies, 77
        labor market, 75, 76
        poverty disparities, 77
        subemployment and unemployment
            rate, 75
    unfavorable conditions
        cost of basic basket, 77–78
        educational establishments, 78–79
        education levels, 77–78
        women, education and labor market, 79
Socio-environmental system, 51–52
Special Law, 51, 92, 95, 110
1998 Special Law
    local leaders, 138–140
    Marine Reserve, 138
    subsidies, 137
Sperm Whales *(Physeter macrocephalus)*,
    229–231
Subsidies, 137
Sustainable energy, 41
Sustainable Tourism Summit, 124


T
Terrestrial ecosystems, 12–13, 188
Tourism
    cluster analysis
        amenities, and health, 114–115
        conservationist, 117–119
        dendrogram, 114
        expansionist, 116
        household characteristics and
            education, 114–115
        isleño and mainland-based lifestyle,
            113, 114
        isolationist, 116–117
    cost of living, 111
    development of, 109–110

economy, 132
ecotourism, 108 (*see also* Ecotourism)
fishing sector, conflicts, 109
"floating hotels," 109
illegal activity, 108
and infrastructure, 29–30
land-based tourism, 109
    advertising strategy, 36
    ecotourism packages, 36
    local residents and politicians, 35
    national (Ecuadorian) tourism, 35
    tourists, accommodation, 35
national economy, 36
*poza de las tintoreras*, 40
shark finning, 40
Special Law, 110
sustainable citizens
    environmental protection, 119–120
    original/native residents, 120
    social and environmental
        irresponsibility, 121
    unlawful activities, 119
tourists, growth rate, 111
urban and rural landscapes, 110


U
United Nations Educational, Scientific and
        Cultural Organization
        (UNESCO), 51
    ecotourism policy, 133
    World Heritage Site, 94
Urban areas. *See* Agricultural and urban areas
Urban site
    carbon concentration, 192, 195
    location, 189
    macroinvertebrate diversity, 196
    nitrogen concentration, 192, 194
    variables, 193
    vegetation cover, 195


V
Volcanic islands, 143


W
Waste recycling, 41
Wastewater management, 176
Water resources
    impacts, 177–178
    impairment, 176–177
    infrastructure, 177

Water source, 152
Whales
    cetacean species diversity, 222–225
    data collection and surveys, 221
    distribution, 225–227
    photo ID studies, 221–222
    wildlife tourism, 220
Whale Sanctuary, 220
Whale watching. *See* Whales

Wildlife tourism, 220
Women of childbearing age (WCA), 81
WorldView-2 (WV-2), 206
World Wildlife Foundation (WWF), 133
World Wildlife Fund, 39, 134

**X**
Xerophytic vegetation, 188